Chipless RFID based on RF Encoding Particle

Remote Identification Beyond RFID Set

coordinated by
Etienne Perret

Chipless RFID based on RF Encoding Particle

Realization, Coding, Reading System

Arnaud Vena
Etienne Perret
Smail Tedjini

First published 2016 in Great Britain and the United States by ISTE Press Ltd and Elsevier Ltd

ISTE Press Ltd
27-37 St George's Road
London SW19 4EU
UK

www.iste.co.uk

Elsevier Ltd
The Boulevard, Langford Lane
Kidlington, Oxford, OX5 1GB
UK

www.elsevier.com

Notices

Knowledge and best practice in this field are constantly changing. As new research and experience broaden our understanding, changes in research methods, professional practices, or medical treatment may become necessary.

Practitioners and researchers must always rely on their own experience and knowledge in evaluating and using any information, methods, compounds, or experiments described herein. In using such information or methods they should be mindful of their own safety and the safety of others, including parties for whom they have a professional responsibility.

To the fullest extent of the law, neither the Publisher nor the authors, contributors, or editors, assume any liability for any injury and/or damage to persons or property as a matter of products liability, negligence or otherwise, or from any use or operation of any methods, products, instructions, or ideas contained in the material herein.

For information on all our publications visit our website at http://store.elsevier.com/

British Library Cataloguing-in-Publication Data
A CIP record for this book is available from the British Library
Library of Congress Cataloging in Publication Data
A catalog record for this book is available from the Library of Congress
ISBN 978-1-78548-107-9

Printed and bound in the UK and US

Contents

Preface

Automatic identification and data capture (AIDC) systems have greatly altered consumers' habits, the monitoring of goods, and services in general. The barcode is the best illustration of this. It was invented in 1949 and has been exploited industrially since 1974. It is still today at the top of the traceability of goods field, where it has been the engine of a distribution market that has become global. Today, the barcode is visible on every purchased product and can even replace in some cases the traditional train ticket or even the movie ticket. The latter can also be used either printed on paper or displayed on a smartphone screen.

In parallel, telecommunication technologies have experienced considerable growth during these last decades. The miniaturization of components has made possible the development of mobile devices with increasingly compact and versatile features. In a few years, we have gone from the bulky landline phone to the smartphone, which allows us to connect to the Internet, to receive many television channels and to navigate in the whole world via the global positioning satellite system. Other developments are underway and will allow the development of reconfigurable and cognitive wireless communicating objects.

In recent years, we have seen the growth of radio frequency identification (RFID) technologies, since manufacturers have begun to pay close attention to what radio frequency waves could bring to the field of identification. Thus, in the urban transport network, we have gone from the paper or magnetic ticket to the contactless card. The improvements provided by this technology are major. For example, passage rates have increased, the maintenance of ticket reading or selling machines has been reduced and the interoperability between different transport networks has been made possible due to the quantity of the onboard information. In addition, logistics platforms seek to gradually replace the barcode with UHF RFID tags for the management and traceability of goods. The main interest is to save time in handling during the control of all objects, which are present on a pallet of goods. This control can be performed without unpacking the palette, remotely and in a fraction of a second. RFID technologies have also brought a lot of new opportunities to the security field. In fact, their performance has opened the way for the reliability of automatic machines and access control.

Despite the benefits brought by RFID technologies, their growth is still restricted by the unit cost of a tag, especially if we compare it to the barcode. In fact, in some applications, the objects to be identified can sometimes have a lower unit price than the price of an RFID tag. It is, therefore, understandable that the conventional RFID technology that uses an antenna connected to a chip cannot be applied in these cases. Thus, in recent years, the study of *chipless RFID* has been of increasing interest and research continues to intensify in this subject. In terms of performance and application, chipless RFID is at the border between the barcode and conventional RFID technology. For this reason, it is sometimes called a "radio frequency barcode". As its name indicates, a chipless tag has no electrical circuit, i.e. no active element and, *a fortiori*, no battery. Therefore, the tag ID is not contained in a non-volatile memory but is directly linked to its geometry in the image of a radar target. In fact, it is the physical structure of the label which, when subjected to an incident electromagnetic wave, will create a suitable electromagnetic signature.

Chipless RFID is a relatively recent technology. We will see that the first published works date back to 2002 and that its application potential is undoubted. However, for technical reasons, there are today very few commercial applications based on its principle. In this book, we will examine how it is possible to remove the technological barriers to allow the development of this new path of RFID technology as an entire identification system. The crucial points for improvement are the increase in the coding capacity, the reduction of the label surface and the possibility to print the tag in order to reduce its unit cost. In parallel, the definition and the design of a reading system (which respects the RF standards) that allows us to detect chipless RFID tags in a robust way is also a blocking point, which will be discussed, as well.

In Chapter 1, a general presentation of the different RFID technologies is given. A brief historical introduction will precede the review of the large families of the systems and applications of RFID technology. We will analyze the strengths and weaknesses of each one.

In Chapter 2, we will focus on the latest developments of the different chipless RFID systems. This will allow us to address the current limitations of this technology and to define the different axes of improvements to consider. To do this, we will place the chipless RFID at the heart of the global identification market.

In Chapter 3, we will address the major issue in chipless RFID, in this case, the coding of information in an RFID tag. In fact, increasing the coding capacities is a major issue that will allow us to impose chipless technology as a real alternative to the current identification technologies: barcodes and conventional RFID technology. One of the objectives is, in particular, to match the coding capacity of EAN13 barcodes. We will introduce performance criteria that will allow us to assess the coding effectiveness of a device according to the occupied bandwidth and the surface required. Different coding techniques will be presented and compared.

In the first section of Chapter 4, we will discuss general information concerning the operating mode of a chipless RFID tag in a spectral signature. An electric model and an analytical model of the basic elements that are contained in the tags will be presented and compared to the results of simulations. Performance criteria, which allow us to evaluate each design, will be introduced. In addition, the design rules, which lead to a particular design taking into consideration the bandwidth and the surface, will be presented and applied on different concepts of chipless RFID tags. The problem of reading chipless tags related to the variability of the surrounding environment will also be addressed and a self-compensation method will be proposed.

The technological barriers related to the low manufacturing cost of chipless tags will be studied in Chapter 5. In the first section of Chapter 5, we will present the manufacturing mode related to the conventional electronics industry, followed by the manufacturing principles of the paper industry. A characterization of potentially usable materials will be proposed, before concluding with a comparison of the performance achieved with these two implementation modes. In the second section, we will present the technical measures developed specifically for the characterization of chipless RFID tags, in a confined space with a metallic cavity, and in an open space using a bistatic radar approach. Two approaches will be explored: the frequency approach using a vector network analyzer and the temporal approach based on the use of an ultra-wideband (UWB) pulse generator and a wideband oscilloscope. The standardization aspect for the UWB communications will be addressed, which will allow us to define the possible detection performance. A reader concept based on the use of a UWB localization radar will be implemented for the detection of chipless RFID tags. To conclude this chapter, signal formatting and decoding will be addressed.

Arnaud VENA
Etienne PERRET
Smail TEDJINI
May 2016

Introduction to RFID Technologies

1.1. Introduction

In this chapter, Radio Frequency IDentification (RFID) technologies will be presented from a general point of view. A brief historical reminder will allow us to return to the applications and the context of the development of the first RFID systems up to the latest advances. The major RFID technologies will be reviewed and grouped according to their operating frequency, their ability to be detected in the near-field or in the far-field and their passive or active nature from an energy point of view. Their performances will be compared and will constitute the reference values for the following chapters, which will focus on the development of chipless RFID.

1.2. The history of RFID

The origin of RFID undoubtedly goes back to the great idea that Brard had in the 1920s to create a form of communication by radiowave between a transmitter (base station, which today is called an "RFID reader") and a device which today is called a "tag" [BRA 24]. The latter is remotely powered by the transmitter. It is composed of a tunable resonant circuit and a switch that allows the modification of the wave which is backscattered by the tag, thus ensuring a modulation of the signal for the communication with the reader. Here, we find the fundamental principle of RFID that still remains at the core of the current systems: a simple tag remotely powered by the reader, with two separate states to establish communication.

Subsequently, the practical implementation of RFID coincided with the World War II and the very important efforts focused on the development of radar. Indeed, it was during this period of unrest that the major technological advances took place and, in particular, RFID [BRO 99]. The British invented the Identification Friend and Foes (IFF) system, a radio frequency transponder system that permitted the identification of allied from enemy planes with the use of encoded signals.

In the former USSR, in 1945, Léon Theremin [GLI 00] invented an espionage system, which was completely passive, allowing it to convert an audio signal into a radio frequency signal with the use of a cavity covered by a diaphragm sensitive to sounds (see Figure 1.1). An antenna is inserted in this cavity, whose volume is altered by the sounds, as shown in Figure 1.1. Thus, the cavity can be considered as a variable load which evolves over time according to the incident sound signal.

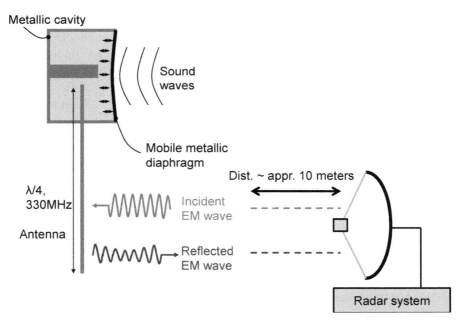

Figure 1.1. *Principle of the listening system "The Thing", invented by Léon Theremin*

The power level of the electromagnetic wave reflected by this antenna will, therefore, vary the rhythm of the sound waves, thus creating an amplitude modulation. This invention may be regarded as the first RFID technology without a chip, even though the identification aspect is not considered in this device. To conclude on this particular period, which is truly the creation of RFID in practice, in 1948, Stockman [STO 48] envisaged the use of what is today known as "RFID" for telecommunication purposes, for various applications using the principle of modulation of the wave power, reflected by a remote transponder. He thus incorporated the principle introduced some years before by Brard. Subsequently, in the 1950s, practical experimentations were developed, along with the patents for the application of this technology [HAR 52, VOG 59].

For almost 20 years, the discipline was explored primarily in the military field, and with the advent of the transistor and the miniaturization of the components, the RFID has become an attractive research discipline [LAN 05]. For example, in 1964, Harrington [HAR 64] introduced the theory concerning the reflection of electromagnetic waves by antennas connected to varying loads.

The first application, which was a commercial success, took place in 1970 with the *electronic article surveillance* anti-theft systems, the transponders of which are the equivalent of tags with a 1-bit chip. Since then, RFID has been the point of interest for large companies such as General Electric and Philips. Other applications emerged, such as the identification of livestock in 1978, with a system marketed by Identronix Research in California. Since the 1980s, research concerning the RFID has not stopped developing. Each application requires specific needs and performances. The constraints vary significantly depending on the operating environment, which explains why the proposed technologies are constantly changing and why research in this area is growing. In the space of 20 years, we have seen the development of electronic toll collection systems to control the access of vehicles on motorways, contactless transport tickets, RFID passports and more recently contactless bank cards. All of these new applications have permanently changed our everyday lives. Even though the principle of chipless RFID was introduced by Léon Theremin in the 1940s, it is only since the 2000s that the research on this very promising subject has started growing. Potentially, chipless technology is expected in the near future to compete with the optical barcode which, up to this moment, is the most used and widespread means of identification.

1.3. RFID technologies

As mentioned above, the wide variety of applications requiring the remote identification of objects explains, in part, the great diversity of RFID technologies that can be found. Thus, the constraints on the reading range, on the nature of the objects to identify (metallic or non-metallic), on the environment of the usage of tags, etc., are particularly diverse. However, the lack of standardization for many years has enabled the emergence and thus the proliferation of competing technologies for the same application.

1.3.1. General operating principle

Despite the incredible number of technological parameters that make up the current RFID systems, the operating principle can be described in a general way. An RFID system is composed of one or several RFID readers, connected or not to supervising computers, which can create a link with databases. These readers allow the identification of objects due to RFID tags that are attached to them, as shown in Figure 1.2.

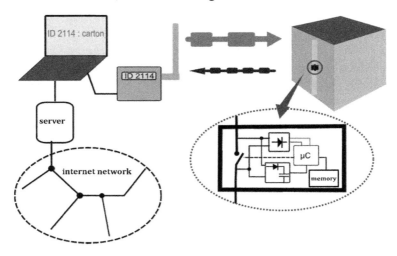

Figure 1.2. *Operating principle of an RFID system*

From an application perspective, we refer to an open system when the tag ID is universal and the association between the ID and the object can be found on a remote server (see Figure 1.2). The barcode is a well-known example of an open system, which allows the identification of articles in a universal way. Closed systems are dedicated to applications in which ID management is carried out internally and is not externally accessible. Thus, the network infrastructure of closed circuits stops at a local network gathering information in a private database. The use of closed systems is justified sometimes by the need for a higher level of security. In this case, IDs must not be known to the public to avoid data tampering.

An RFID tag is composed of an antenna designed to operate in a given frequency band, connected to an electronic chip. A matching circuit is necessary in some cases to match the impedance of the antenna to that of the chip.

In an electronic chip, we generally find:

− an energy harvester circuit achieved with the aid of a diode rectifier;

− an asynchronous demodulator or a diode envelope detector for the reception of queries from the reader;

− a microcontroller for the processing of queries, cryptography and the preparation of responses;

− a memory that can contain the tag ID and other information specific to the application;

− an electronic switching circuit, which allows the modulation of the complex impedance of the tag in order to generate a response.

RFID tags can be classified according to their power supply mode, their operating frequency, their cryptographic capability, their communication protocol or even by the presence or absence of an electronic chip.

Thus, we refer to passive tags when the tag is completely powered by the electromagnetic field of the reader. Semi-passive tags use the energy of the reader to generate the response to a query from the reader. In contrast, the other elements of the chip such as the microcontroller and the memory draw their energy from a battery. Finally, an active tag is completely powered by a battery. It generates the response to the reader from its own energy using a radio frequency frontend. In this last case, the transmitting and receiving frequencies may be different. According to the operating frequencies, the physical principles involved are different, which leads to very different performances in terms of reading range, tag positioning, unit cost and susceptibility to the environment (see Table 1.1).

Family	Range	Coding capacity	Access	Level of confidentiality	Susceptibility	Positioning	Tag cost
LF, HF	<1 m	Some kbits	Read/write	High	Metal at 13.56 MHz	Polarization independent	>0.4 euros
UHF, SHF	1-100 m	Some kbits	Read/write	Low	Metal and liquids	Polarization dependent	>0.1 euros
UWB	<60 m	Some bits	Read/write	Good	Metal and liquids	Polarization dependent	>0.3 euros
Chipless	<1 m	256 bits	Read/write	None	Metal and liquids	Polarization dependent	>0.005 euros

Table 1.1. *Classification of RFID technologies according to usage frequencies*

Thus, we can classify the RFID technologies in four subfamilies:

– the Low Frequency (LF), High Frequency (HF) technologies in a magnetic coupling;

– the Ultra High Frequency (UHF), Supra High Frequency (SHF) technologies which use the propagation of electromagnetic waves;

– the Ultra-WideBand (UWB) technologies;

– the chipless technologies.

The reading range is related to the operation mode of the tag in the near-field or in the far-field, as shown in Figure 1.3. LF tags operate in the near-field. Their operating distance is much less than the wavelength λ, equal to 2,400 m at 125 kHz and 22 m at 13.56 MHz. The energy transfer is mainly carried out by an inductive coupling. For the UHF-SHF frequencies, the wavelength is between 0.7 m (433 MHz) and 5 cm (5.8 GHz). Waves propagate from the transmitting antenna, along a distance R that can be obtained by relationship [1.1]. In this whole area, the approximation in the far-field can be used. In this equation, D represents the largest dimension of the transmitting antenna. For example, at 915 MHz, for a maximum antenna dimension D of 15 cm, the area of the far-field is beyond $R = 14$ cm.

$$R > 2 \cdot \frac{D^2}{\lambda} \qquad\qquad [1.1]$$

Tags in these frequencies, therefore, operate mainly in the far-field regime. The attenuation of the electric field no longer follows the $1/R^3$ rule as in the near-field, but the $1/R$ rule for a spherical wave, thus favoring an energy transfer for greater distances.

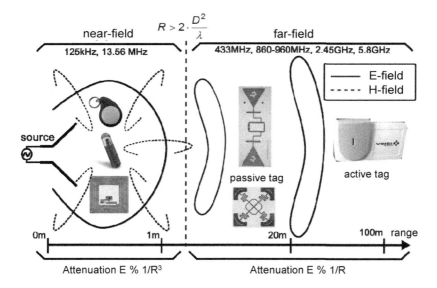

Figure 1.3. *Classification of RFID technologies according to their area of operation (near-field and far-field) and their reading range*

1.3.2. *LF and HF technologies*

"Near-field" technologies are mainly found in applications where confidentiality of data exchanged between the tag and the reader must be guaranteed. Take, for example, the passport, or even the RFID ticket. The readers used in this case have a reading range of about 10 cm. The data exchange can be performed in an encrypted mode, for example, when it comes to handling money (the amount of a transport unit). This means that the chips implemented on this kind of tag are actually microcontrollers with sufficient memory and calculation units dedicated to cryptography. Communication rates can reach 848 kbit/s, which allows the fast reading of a significant amount of information. In the case of a passport, the identity photo of the holder can be read in a fraction of a second. The operating frequencies mainly used are 125 kHz, 134 kHz and 13.56 MHz. They belong to the industrial, scientific and medical (ISM) bands. The 125 and 134 kHz bands are reserved for applications requiring the storage of a small amount of information, and they have the advantage of a better tolerance to metal environments.

The operating principle of an RFID tag in the near-field [FIN 10] is described in Figures 1.2 and 1.4. The reader generates a continuous wave (CW), the carrier wave, for example at 13.56 MHz. The remote tag is powered by magnetic coupling, like a transformer. The energy collected by the tag is recovered through a diode bridge and powers the chip. To wake up a remote tag, the reader sends a command by modulating the carrier wave in amplitude with a modulation index of 100%. At the tag level, a simple diode detector allows the demodulation of the query from the reader. After the analysis of the query by the chip, the response of the tag is sent using the principle of load modulation.

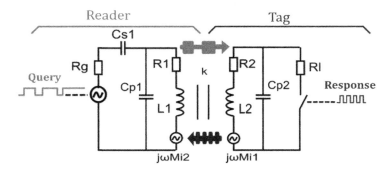

Figure 1.4. *Operating principle of an RFID tag in a magnetic coupling*

To better understand the procedures that take place at the reader level during the charging of the tag module, we can make the analogy of the operation of an electrical transformer (see Figure 1.4) in which the secondary winding is connected to the primary winding by a magnetic coupling and vice versa [FIN 10, VEN 09]. If the secondary winding of the transformer does not have a load, the secondary current is zero, which induces no opposite electromotive voltage in the primary winding. However, an electromotive force at the primary winding proportional to the level of the current in the secondary winding will reduce the total voltage at the primary winding. Thus, the current modulations at the secondary winding are, therefore, detectable at the primary winding. This is the principle used in the RFID systems in a magnetic coupling. Thus, by modulating the load at the tag level, at the rhythm of a frequency submultiple of the frequency of the carrier wave (generated by the division of the frequency of the carrier wave, 13.56 MHz), the current is modulated in the loop antenna of the tag, and by magnetic coupling, the voltage at the loop antenna level of the reader is also modulated.

The standard governing the operation of RFID tags at 13.56 MHz from the transport layer up to the application layer is ISO/IEC 14443 [FIN 10]. This standard applies worldwide. It has allowed the unification of the development of RFID systems at 13.56 MHz and has, therefore, made their use widespread in the field of transport, access control to buildings, passports and recently contactless payment methods with Paypass. Many manufacturers propose the operation of the products in the 13.56 MHz band. The coding capacities of these chips are generally between 256 bits and several tens of Kbits.

The LF systems operating mainly at 125 and 134 kHz are more heterogeneous and not part of the ISM band. Their main applications are in access control to buildings (gates) or micropayments for automatic beverage dispensing machines. For the purposes of livestock traceability and in order to fight against animal trafficking, pets are also tagged with chips at 125 or 134 kHz.

1.3.3. *UHF and SHF technologies*

RFID technologies in the UHF and SHF bands have emerged in the last decade and are generating a real interest, in particular since the definition of the Electronic Product Code (EPC) standard by the Auto-ID Center in 2003. The EPC brings together a consortium of 120 major companies in the field of identification and RFID technologies. To name just a few, the intended applications concern the traceability of goods or the management of pallets in logistics centers. The EPC standard has been adopted by the ISO for the definition of the ISO 18000-6 C standard, dedicated to the so-called second-generation chips operating in the 860–960 MHz frequency band.

Even before the creation of the EPC, UHF RFID systems have been used at the French motorway tolls since 1992. In fact, the need to have greater reading distances than those possible with the technologies based on the magnetic coupling motivated the development of this technology for the 2.45 and 5.8 GHz frequencies. These two frequencies are subject to the ISO 18000-4 and 18000-5 standards. Finally, more recently, the usage of the 433 MHz frequency was the subject of a new addition to the ISO 18000-7 standard.

Hereinafer, we present two main technological variants: passive tags and active tags. They are differentiated by the way in which the response is returned to the base station. We refer to passive tags when the response of the tag to the reader is based on a backscattering principle. In contrast, an active tag embeds a real radio frequency transmitter. Therefore, its transmitting power is not related to the distance between the tag and the reader, because the energy is provided locally by a battery. The choice between a passive and an active tag can be made according to several criteria, including the reading range, the tag cost, the size, the lifetime, etc.

1.3.3.1. Passive tags

In recent years, passive tags are the most widespread tags and they exibit the largest growth in terms of units sold. The most used frequencies are in the 860 and 960 MHz band. At these frequencies, the preferred operation mode of these RFID systems is the far-field. In this way, greater reading ranges can be achieved. A UHF/SHF tag is composed most of the time of a dipole antenna, which allows us to capture the electromagnetic radiation. This antenna is designed to have an impedance matched to that of the chip (exactly in the state of impedance in idle state), which is directly connected to its terminals. The diagram of the general principle of an RFID system presented in Figure 1.2 remains valid. In the same way that the HF or LF RFID tags are presented above, passive tags operate without a battery. Thus, it is the electromagnetic field of the base station that remotely powers the chip of the tag. The latter contains an energy harvester circuit, which powers a logic circuit and a memory accessible in read/write mode. Typically, a voltage doubler using diodes followed by a capacitor is used for the rectification of the carrier wave. A variable resistor in parallel to the capacitor allows the limitation of the voltage when the tag gets too close to the base station. According to the commands sent by the base station, a read access (most of the time) or a write access can take place. The response of the tag is then transmitted by incorporating the principle of load modulation shown in Figure 1.5.

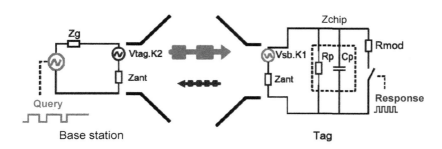

Figure 1.5. *Operating principle of a UHF/SHF passive tag: load modulation*

The interaction between the tag and the reader is modeled by voltage sources. At the tag level, the Vsb·K1 value represents the terminal voltage of the radiation resistance of the reader antenna (Vsb) multiplied by a K1 factor, which takes into account the free space attenuation, the radiation efficiency, the polarization factor, the losses due to mismatching and the antenna gains of the reader and the tag. When using the far-field approximation, this K1 factor decreases by $1/R$ as a function of distance. In the same way, the interaction of the tag on the reader is modeled by a voltage source of the Vtag·K2 value. "Vtag" represents the terminal voltage of the radiation resistance of the tag antenna. This voltage is a function of the voltage captured by the tag (Vsb·K1), and of the chip impedance state. "K2" is equal to "K1" if the antenna at the reader level is used both in transmission and in reception (antenna reciprocity principle). It is interesting to quantify these terms in order to establish the link budget and to estimate the reading range. For reasons of convenience, from this point forward, we will refer to power instead of voltage.

We have considered that the essential criterion defining the reading range is the power received at the chip level. Similarly, we have considered so far an optimal energy transfer to the chip. Let us look more closely at the matching problems between the chip and the antenna. Figure 1.6 represents the equivalent circuit that models the energy transfer between the chip and the antenna in the case of matching (Figure 1.6(a)) and modulation by a short-circuit (Figure 1.6(b)). In order to maximize the energy transfer of the tag antenna to the chip (or the load), a complex conjugated matching should be performed [1.2]–[1.3]:

$$Zant = Rant + jXant, \quad Zchip = Rchip + Xchip, respectivelly \qquad [1.2]$$

$$Rant = Rchip, \quad Xant = -Xchip \qquad\qquad\qquad [1.3]$$

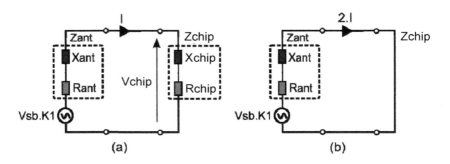

Figure 1.6. *a) Matching of the antenna to the chip in unmodulated mode; b) equivalent circuit when the chip is modulated (short-circuit in parallel – Rmod = 0). The current in Rant is two times more important*

In the case of a complex conjugated matching, half of the energy is dissipated in the chip, the other half is dissipated in the resistance of the antenna, which in general is composed of two parts, a radiation resistance and a resistance that models the losses in the antenna (joule loss and dielectric loss).

When the chip modulates the carrier wave to generate the response to the reader, in a first approximation, we can consider that the input impedance of the chip is loaded by a transistor operating in blocked/saturated mode. When the transistor is saturated, there is no more energy transfer to the chip and the power captured is completely backscattered in the open space. The current flowing in *Rant* is 2 times higher than the case of a perfect matching, which translates into a 4 times greater backscattering power in the case of a matching [1.4] (see Figure 1.6(b)). In fact, in the case of a passive tag, to avoid deenergizing the chip in this configuration, we use a low impedance state rather than a short-circuit.

$$Ptag_{match} = Rant \cdot I^2 \,, \; Ptag_{cc} = Rant \cdot (2I)^2 = 4 \cdot Rant \cdot I^2 = 4 \cdot Ptag_{match} \quad [1.4]$$

The principle of response generation of the tag to the reader is, therefore, based on the variation of the backscattering power between two quite distinct states. A logic state "1" is characterized by a continuous wave reflected with an amplitude of up to 4 times greater than a logic state "0". The base station, therefore, detects a signal modulated in amplitude, whose variation depends directly on the variation of the impedance state. For non-battery-assisted passive tags, we avoid in general the short-circuit state in the strict sense, to maintain a minimum of energy transferred to the chip during the response. Thus, the Rmod resistance in Figure 1.5 can have a value of approximately a hundred ohms.

The tag antenna is a key element that directly affects the reading range in the same way as the activation power of the chips. The important parameters to evaluate during the design of an antenna of an RFID tag are:

– the efficiency;

– the matching;

– the radiation pattern;

– the variation of the radar cross-section (RCS) between the two impedance states (Delta RCS).

The antenna design must, therefore, take into account these parameters, the size constraints and the environment of use. In most cases, these impose performance limits [PER 12, PER 14]. The increase in the reading range of the tag is a function of the antenna gain. However, a high gain is hardly feasible in compact dimensions. In addition, increasing the gain is achieved at the cost of the antenna aperture, which risks becoming too directive and, therefore, hardly detectable according to the orientation, and the positioning of the tag. In practice, most of the time, we search for an omni-directional behavior. As previously stated, a condition of complex conjugated matching must be carried out between the antenna and the chip. In practice, for cost reasons, the matching is carried out at the antenna level without using a matching circuit, so that it can be connected directly to the chip. Finally, the radiation pattern of the antenna should respect the constraints of the tag usage. In general, antennas of RFID tags have an isotropic radiation in the H plan, to ensure their detection regardless of the direction of the incident field.

To establish a link budget, we can use Friis equation [1.5] that takes into account the characteristics of the transmitter, the receiver and the free space attenuation as a function of the operating frequency and distance, everything in the far-field approximation.

$$P_{rx} = P_{tx} \cdot \left(\frac{\lambda}{4\pi R} \right)^2 G_{rx} G_{tx}, \qquad \lambda = \frac{c}{f} = \frac{3 \cdot 10^8}{f} \qquad [1.5]$$

$$P_{rx} = P_{tx\,eirp\,max} \cdot \left(\frac{\lambda}{4\pi R} \right)^2 G_{rx} \qquad [1.6]$$

P_{rx} and P_{tx} represent the powers at the receiver and transmitter level, and G_{rx} and G_{tx} represent the gains achieved by the receiving and transmitting antennas, which take into account the potential losses due to mismatching. R represents the distance of the reader/tag, and λ is the wavelength in free space. Equation [1.5] can be reformulated in [1.6] in order to show the maximum equivalent isotropically radiated power $P_{txeirpmax}$ which is a regulated figure and whose value depends on the used ISM band (see Table 1.2). In a passive tag remotely powered by the electromagnetic field of the reader, the criterion that limits the detection range of the tag is primarily linked to the minimum activation power of the chip [PER 12]. The receiver at the reader level is generally very sensitive, as it is more evolved and has an energy source. The reason is primarily technological. A silicon chip needs voltage and a minimum current to operate its internal logic (transistors, diodes, etc.). For example, Impinj Monza 4 (IMP) chips can operate in the 860–960 MHz frequency range with a minimum power of P_{rx} min = –20 dBm (10 µW) (read) and –16 dBm (25 µW) (write). We can thus draw a table comparing the reading range of passive RFID tags in the case of a perfect matching of the tag antenna with the chip, by taking into account the maximum permitted transmitting powers. Starting from equation [1.6], we can express R [1.7] as a function of $P_{rx\,min}$, which corresponds to the activation power of the chip, of $P_{tx\,eirp\,max}$ and of the receiving antenna gain for a given frequency. Equation [1.8] allows us to calculate R based on

$P_{txerp\ max}$, the radiated power relative to a dipole $\lambda/2$. In this expression, 1.64 represents the directivity of the dipole $\lambda/2$.

$$R = \frac{\lambda}{4\pi\sqrt{\dfrac{P_{rx\min}}{P_{txeirp\max} \cdot G_{rx}}}} \qquad [1.7]$$

$$R = \frac{\lambda}{4\pi\sqrt{\dfrac{P_{rx\min}}{P_{txerp\max} \cdot 1.64 \cdot G_{rx}}}} \qquad [1.8]$$

The reading ranges presented in Table 1.2 are theoretical and represent a maximum value, which can be achieved in the ideal case (perfect matching between the chip and the antenna, no reflection, not a noisy environment, etc.). In a practical operating case, i.e. in actual conditions, these values can easily be divided by a factor of several units [PAR 09].

Frequency band (MHz)	Region	Power	NXP UCODE HSL (NXP) Ptag = –14.5 dBm @ 860 MHz and –9.2 dBm @ 2.45 GHz	NXP UCODE Gi2L (NXP) Ptag = –18 dBm	Impinj Monza 4 (IMP) Ptag = –20 dBm
869.4–869.65	Europe	0.5 W ERP	5.38 m	8 m	10 m
865.5–867.6	Europe	2 W ERP	10.75 m	16 m	20 m
902–928	USA	4 W EIRP	11.28 m	16.9 m	21.2 m
2,400–2,483.5	Europe	0.5 W EIRP	0.8 m	/	/
2,400–2,483.5	Europe/USA	4 W EIRP	2.27 m	/	/

Table 1.2. *Theoretical reading ranges for different chips according to usage frequencies*

When more important reading ranges are required, we can add a battery at the tag level to supply power only to the digital part (microcontroller and memory). The modulation stage remains the same. The backmodulation principle is, therefore, always used. Sometimes, we call this variant "battery-assisted passive tag" or even "semi-passive tag". To determine the reading range, we must take into account in this case the downlink (reader to tag) and the uplink (tag to reader). It is possible in this case that the tag is activated by the reader, but that the backscattered signal is too low for the reader to be able to interpret it. In order to determine the detection ranges in this case, it may be interesting to use the RCS of the antenna, which is a parameter that is linked to its geometry on the one hand (structural mode), and to its load (antenna mode) on the other hand [HAR 64]. RCS represents the area equivalent to the area which would capture the density of the power transmitted by a source located in the far-field at a distance R and which would retransmit it in the entire space [BAL 05]. Under far-field approximation, the expression of the RCS is given by [1.9].

$$\sigma = \lim_{R \to \infty} \left[4\pi R^2 \frac{|W_s|}{|W_i|} \right] \qquad [1.9]$$

In this equation, W_i represents the power density that arrives at the target (or the tag) level. This power density is transmitted by a source located at a distance R. W_s is the power density backscattered by the target in a given direction. The far-field condition is valid if [1.1] is verified. In order to perform a power budget as a function of the RCS, we can use the radar equation [1.10], which allows us to calculate the reflected power at the source level by a target located at a distance R. G_{tx} and G_{rx} are the receiving and transmitting antenna gains, and σ represents the RCS in meters.

$$\frac{P'_{rx}}{P_{tx}} = \frac{G_{tx}G_{rx}\lambda^2}{(4\pi)^3 R^4} \sigma \qquad [1.10]$$

The RCS value is related to the tag antenna gain in a given direction. Formula [1.11] allows us to make the link between the antenna gain and RCS when the load is matched. Equation [1.12] allows us to obtain the RCS value when the load is in short-circuit. Thus, we note that the ratio of 4 between these two load impedance states is already highlighted by the relationship [1.4].

$$\sigma_{matched} = \frac{\lambda^2}{4\pi} Gant^2 \qquad [1.11]$$

$$\sigma_{cc} = \frac{\lambda^2}{\pi} Gant^2 \tag{1.12}$$

In order to estimate R, we must take into account the variation of the minimum detectable power by the reception stage of the reader. With the radar equation [1.10], we can link this power variation to a variation of RCS between the two states generated by the remote tag, also known as Delta RCS and noted here as ΔRCS. Thus, equation [1.10] can be put in the form [1.13] and the reading distance can be obtained by reformulating [1.13] to [1.14].

$$\Delta P_{rx} = P_{tx} \frac{G_{tx} G_{rx} \lambda^2}{(4\pi)^3 R^4} \Delta\sigma \tag{1.13}$$

$$R = \sqrt[4]{\frac{P_{tx\,max} G_{tx} G_{rx} \lambda^2}{\Delta P_{rx\,min} (4\pi)^3} \Delta\sigma} \tag{1.14}$$

In this case, the limiting factor is the sensitivity threshold of the receiver that is typically -80 dBm [NIK 06] for tag to reader link. A value of -70 dBm is preferable in order to ensure a fairly low error rate. In this case, because the electronics of the tag are locally powered, its activation power is much lower. The uplink (reader to tag) is, therefore, no longer the limiting factor of the reading range. In Table 1.3, we provide some theoretical ranges calculated with [1.14] based on the sensitivity parameters set out above. UHF bands typically used in RFID are between 860 and 960 MHz, as well as 2.4 GHz and the transmitting powers are in agreement with the regulatory authorities in Europe and the United States. In order to characterize the Delta RCS of conventional tags, we have taken first the analytical formula of a dipole whose load varies between the matched case and the short-circuit case. With equations [1.11] and [1.12] and taking the gain of a dipole which equals 1.64, we can deduce expression [1.15].

$$\Delta\sigma = \sigma_{cc} - \sigma_{matched} = \frac{\lambda^2}{\pi} 1.64^2 (1 - \frac{1}{4}) = 0.642\lambda^2 \tag{1.15}$$

This value is finally compared to the reference value commonly used in UHF RFID, which sets a minimum Delta RCS of 50 cm^2.

Frequency band (MHz)	Region	Power	Dipole ΔRCS λ/2: 0.642· λ²	Min. ΔRCS: 0.005 m²
869.4–869.65	Europe	0.5 W ERP	10 m	7 m
865.5–867.6	Europe	2 W ERP	19.7 m	10 m
902–928	USA	4 W EIRP	19.8 m	10.2 m
2,400–2483.5	Europe	0.5 W EIRP	4.3 m	3.7 m
2,400–2,483.5	Europe/USA	4 W EIRP	7.3 m	6.2 m

Table 1.3. *Theoretical reading ranges for battery-assisted passive tags for different ΔRCS values, (ΔPrxmin = –70 dBm) calculated with [1.14]*

1.3.3.2. *Active tags*

When the required reading ranges are greater than 10 m, the use of active RFID tags may become more appropriate. Even reading ranges of 50–100 m are possible. The most popular application of active RFID is the access control of vehicles at the motorway tolls. What is of interest is to be able to detect the vehicles approaching the toll gate without them having to stop. A traveling speed of 30 km/h is possible. As we have already mentioned, "active" means that the tag embeds a radio frequency transmission module. In this case, the base station no longer needs to maintain the electromagnetic field during the response of the tag. The tag can also respond on a frequency other than that of the base station. This makes possible a "full duplex" communication mode. A direct communication between two tags can also take place as shown in Figure 1.7. An active tag embeds, therefore, a battery to power both its digital electronic and its transmitter, in the same way as in a general-purpose radio transmitter (see Figure 1.7). In order to reduce the power consumption of the battery, a wake-up procedure can be implemented. The lifetime of the tag is often related to the lifetime of the embedded battery and can go up to 7 years (for example, Savi Tags [SAV 16]). Commonly used frequencies are in the 433 MHz band, as well as in the 2.45 and 5.8 GHz bands. Active RFID technology is primarily used at motorway tolls, and also for the management of containers, and anti-theft systems for high added value products. The calculation of the transmission/reception range between the tag and the base station or another tag uses Friis equation [1.2]–[1.5]. The detection distance in this case is limited by the transmitting power of the active tag of 1–10 mW. This limit is primarily determined by the size constraints, the cost and the transmitter autonomy. These constraints are often more important than the regulatory limits on radiated power. In Table 1.4, we have gathered theoretical range values for tags by applying Friis formula for the "tag to base station" link. The reception sensitivity used is −70 dBm.

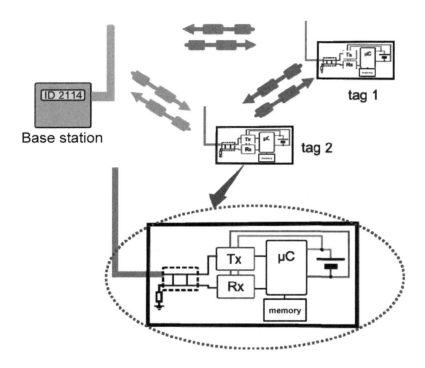

Figure 1.7. *Communication diagram of an active RFID tag. Communication between the base station and the tags or directly between tags*

Frequency band (MHz)	Theoretical range (1 mW ERP)	Theoretical range (10 mW ERP)
433	285 m	900 m
869	142 m	450 m
2450	50 m	160 m

Table 1.4. *Theoretical reading ranges for active tags (Prxmin = −70 dBm)*

1.3.4. *Ultra-wide-band technology*

UWB RFID is a recent discipline that represents an interesting alternative to narrowband RFID systems which were described previously. What has motivated research teams and manufacturers to explore this technology is primarily the possibility to precisely locate tags, and also to identify them. In fact, in UWB communication, ultrashort pulses (of the order of a nanosecond) are used to convey the data, which allows us to estimate with precision the duration of a two-way journey or the time of flight performed by the signals. This also allows us to choose in a very precise way the detection zone of the tags if the reader antenna is capable to dynamically modify its radiation pattern.

However, using signals with a very wide frequency spectrum allows us to overcome interference problems that arise with narrowband communications. A spectral spread is more robust in different signal attenuation mechanisms and in multi-path environments. This promotes a more significant reading range, at equal power, in comparison to a narrowband RFID system.

A UWB communicating system is characterized by a bandwidth greater than or equal to 500 MHz. In the United States, the Federal Communications Commission (FCC) authorizes UWB communications between 3.1 and 10.6 GHz. In Europe, the Electronic Communications Committee authorizes communications between 3.1 and 9 GHz with an unauthorized band between 4.8 and 6 GHz. Figure 1.8 illustrates the different authorized transmission masks. The maximum authorized power spectral density is −41.3 dBm. These values may seem low but in impulse radio, duty cycle and the duration of exchanged signals is low. It is, therefore, preferable to refer to energy transported by a pulse. The calculations of detection range, therefore, use energies rather than powers.

Reading rates of 10 Mbit/s are achievable with UWB RFID systems, while guaranteeing a good reading range. A direct relationship can be established between the detection range and the data rate. In fact, in the standard, pulse repetition frequency per second (PRF) defines bit rate, as well as the average power required for their transmission. For example, at equivalent average power, a PRF of 1 MHz will allow us to transmit 10 times more energy for each pulse than with a PRF of 10 MHz.

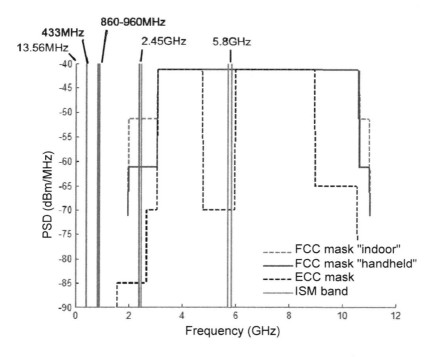

Figure 1.8. *UWB communication mask*

The hardware architecture of a UWB RFID reader is simpler for the transmission part. The pulse generator is directly connected to the antenna without need for a mixer and a local oscillator. This provides a better energy efficiency than with a narrowband transmission [ZHU 11].

The reception part at the tag and the reader level is more difficult to achieve and we distinguish several approaches [ZHU 11]. The first is a direct approach, in which an analog-to-digital converter is connected to the receiving antenna via an amplification stage. The processing is carried out in this case digitally. This approach may prove to be complex given the computing power, which is necessary to process the flow of samples for frequency signals between 3.1 and 10.6 GHz. This leads to the use of converters with a sampling frequency greater than twice the maximum frequency (Shannon theorem). Other approaches use analog correlators to compare the received signal with a reference pulse. A very good sensitivity can be achieved with this kind of reception stage, but at the expense of synchronization and a significant power consumption. Asynchronous detectors are less sensitive but simpler to implement and, in particular, they have a low consumption. They are based on the detection of an energy level on a given period. We use a square law detector followed by an integrator in finite time whose integration period corresponds to the duration of a bit.

UWB RFID tags can be active or passive [GUI 10, DAR 10, DAR 08]. Passive systems use a load modulation principle as in the conventional RFID. Commonly used codings are pulse position modulation (PPM), pulse amplitude modulation (PAM) or on off keying (OOK). Figure 1.9 presents the possible architecture of a passive UWB tag [DAR 08] as a function of the type of modulation used.

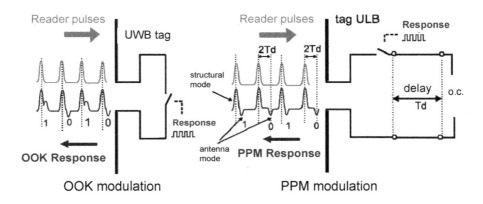

Figure 1.9. *Architecture of a passive UWB tag*

In a passive system, the UWB reader sends pulses at regular intervals. These pulses are captured by the remote tag, as shown in Figure 1.9. In the case of a PPM coding, the pulse is retransmitted directly or via a delay line, switched depending on the binary state to be transmitted. In the case of an OOK or PAM coding, the pulse can be reflected via a short-circuit (or an open circuit), which provides a phase difference of 180° (or 0°) between the two reflected states.

In reality, we need to define more precisely the mechanism of tag backscattering in the case of UWB signals. Irrespectively of the load connected to its terminals, a signal is in all cases reflected by the tag antenna. This reflection mode is structural. It depends only on the geometry of the antenna. Another part of the reflection known as antenna mode is linked to the load connected to the antenna. It is, therefore, this reflection mode that varies depending on the configuration of the load (see Figure 1.9). For UWB antennas, the reflection linked to the structural mode is even preponderant compared to the antenna mode.

UWB semi-passive tags provide a good performance in terms of the reading range. Remember that "semi-passive" means that a principle of passive backmodulation is used to generate the response. The digital part of the tag is powered by a battery. Thus, it is reported that a distance of 20 m can be reached for a rate of 1 kbits/s with an average transmitting power of 0.09 mW [DAR 10].

Commercial solutions of UWB RFID systems today are all based on the use of active tags. In most cases, the tag is powered by a battery [ZEB 16, DEC 16]. The active tag "DartTag" by Zebra Technologies [ZEB 16] allows detection at a distance of up to 100 m in free space with a reading rate of 3,500 tags/s. The autonomy of tags is approximately 7 years at a rate of one reading per second. In addition, it is possible to locate a tag with an accuracy of 30 cm. Regarding the "DW1000" tag by decaWave [DEC 16], it is reported that a reading range of 450 m in free space and of 45 m indoors can be achieved, by ensuring data transfer at 6.8 Mbit/s. The positioning of a tag within 10 cm distance is possible.

To calculate the reading ranges of a UWB tag, we can use Friis equation [1.4] and get relationship [1.16] [ZHU 11].

$$R = \frac{\lambda}{4\pi \sqrt{\dfrac{N_0 \cdot N_F \cdot PI \cdot SNR_{bit} \cdot PRF}{DSP_{eirp\max} \cdot BP \cdot G_{rx}}}} \hspace{2cm} [1.16]$$

The frequency bandwidth used, the noise power related to the power spectral density of white noise N_0 and the noise factor N_F of the reception stage (low-noise amplifier) will define the sensitivity threshold. The theoretical signal-to-noise ratio which is required at the level of SNR_{bit} bit is a function of the desired *bit error rate* (BER) and the coding used. For example, for an OOK coding, and a BER of 10^{-3}, an SNR_{bit} of 12.5 dB is necessary [ZHU 11]. Implementation losses *PI* can be added to take into account the imperfections of the detection stage. The transmitting power corresponds to the authorized power spectral density $DSP_{eirpmax}$ multiplied by the bandwidth BP of the pulse. The parameters that can be modified to improve the reading distance are the receiving antenna gain G_{rx}, the pulse repetition frequency PRF, as well as the pulse bandwidth BP. It should, however, be noted that the standards impose a limit on the maximum energy contained in a pulse. The FCC defined a minimum PRF of 1 MHz for a maximum BP of 7,500 MHz. Table 1.5 presents the theoretical range values for several values of PRF and BP, which are calculated according to the FCC standards. The power spectral density is defined at –41.3 dBm on the entire UWB band between 3.1 and 10.6 GHz. The parameters IF and NF correspond to the reception stage defined in [ZHU 11].

Bandwidth (MHz)	Theoretical range (PRF=1 MHz)	Theoretical range (PRF=10 MHz)	Theoretical range (PRF=100 MHz)
500	29 m	9.2 m	2.9 m
2,000	47 m	15 m	4.7 m
7,500	55 m	17.4 m	5.5 m

Table 1.5. *Theoretical reading ranges for active UWB tags using the reception stage defined in [ZHU 11] (N0 = –174 dBm/Hz, NF = 10 dB, PI = 5 dB, SNRbit = 12.5 dB, Grx = 1)*

However, there are solutions where the tag is remotely powered [TAG 16, MUC 08, REU 06] and in this case, the diagram differs slightly (see Figures 1.10(a) and (b)). In the *Tagent* solution, to power the tag, we use a source of an electromagnetic field of UHF or SHF frequency contained in an ISM band. We use beacons of remote powering near the tags (max. 1 m) to power them with a CW wave at 5.8 GHz. The tag stores the energy to power its electronics and to transmit responses to the base station in the form of UWB pulses. With this process, a detection range of 10 m is possible, on condition that the tag is close to a remote powering source. The chip developed by *Tagent* integrates an antenna in an area of 2.3 × 2.3 mm². This reduces the connection problems between the antenna and the chip during the encapsulation of the tag with its support (ticket, etc.). A tag unit cost of 0.3$ is reported, making it quite a competitive solution with a more considerable investment on the reading system (2,000$ for the reader, 50$ per beacon of remote powering) in comparison to a conventional RFID system. Other solutions [MUC 08, REU 06] use a UWB link both to power the tag and to communicate in both directions (see Figure 1.10(b)). Unlike a conventional RFID system, the transmission phase of a command or the reception phase of the response is separated from the power phase of the tag. To power the tag, the reader sends a series of pulses at a constant interval. The energy harvester device is a charge pump that stores the energy in a capacitor. An alternation between the energy harvesting phase and the data transmission takes place.

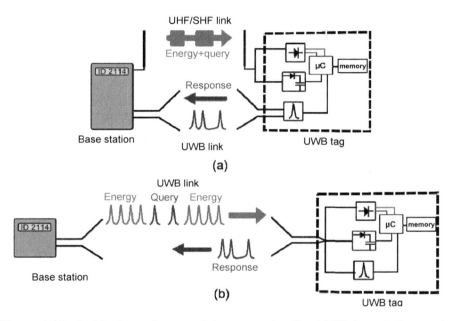

Figure 1.10. *Architecture of a remotely powered active UWB tag: a) the reader commands and the energy are provided via a narrowband UHF/SHF link. The response of the tag is in UWB; b) the communication and the energy of the reader to the tag are provided via the UWB link. The response of the tag is in UWB*

Even though it is interesting for the performance achieved in terms of range (active tags) and positioning, UWB technology is still not quite widespread. The reasons are the strictness of the regulation standards, especially in Europe, regarding the use of frequency covering the 3.1–10.6 GHz band, as well as the relatively few years that "impulse radio" discipline has been used.

1.4. Conclusion

This section has allowed us to introduce the different families of RFID technologies. Depending on the applications, we can have different solutions. When a reading range greater than 1 m is deemed necessary, UHF and SHF tags are used, while in the opposite case, HF tags are omnipresent in particular for access control applications. Passive or semi-passive technologies are mostly used contrary to the active technologies. UWB RFID technologies are still not widely used even if they have a better performance in terms of speed, data transfer and positioning. The predominantly used technology today is HF technology, in particular in the field of transportation where it has been successfully imposed. UHF technologies are mostly found in logistics platform to increase the operating speed and the loading procedures. These technologies are successfully used when the tag cost becomes negligible compared to the value of the article that it identifies. For example, a subscription to a transportation program is relatively expensive (30–200 euros/month) compared to the price of a contactless card (40 cents). The price of a UHF tag of 10 cents is also much lower than the value of a palette of goods. However, this limits the mass deployment of conventional RFID tags, when we seek to identify low-cost products with high-power consumption compared to the tag price. With this observation, the idea of creating chipless RFID tags became obvious. We will see in the following chapters that chipless RFID technologies have strong arguments in order to address these mass markets. In fact, low-cost tags (less than a euro cent) can be made on paper or plastic substrates. Reading ranges of 1 m are possible, with detection systems that approach those developed for UWB RFID. In addition, owing to the absence of chip, this technology can be quite robust at the mechanical level. It is able, for example, to operate in extremely severe environments of use, which are subjected to strong temperature variations, intense radiation or even electrostatic discharges.

2

The Latest Developments on Chipless RFID Technologies

2.1. Introduction

The objective here is to present the latest developments of chipless RFID technologies. A classification of different tags is proposed, by presenting their performance in terms of detection range, coding capacity and size requirement. The results of market studies on the development of different chipless RFID technologies will be discussed. A report of the different current limitations of chipless technologies will allow us to establish the issues explored in this book.

2.2. The latest developments on chipless RFID technologies

Chipless RFID is a recent research discipline. The first articles mentioning the design of chipless tags emerged in 1988 [NYS 88] with surface acoustic wave tags, also known as SAW [HAR 02], or even with tags based on multi-resonant structures at a few tens of MHz, which were compatible with low-cost manufacturing technologies [FLE 02]. However, it is remarkable that such a design was introduced more than half a century ago with one of the first of all RFID systems invented by Léon Theremin in 1945. This device, as described in Chapter 1, was probably the first chipless RFID system.

The different RFID technologies introduced earlier may seem very different from each other. Yet, in all cases, we find two key elements constituting an RFID tag, i.e. an electronic chip connected to an antenna.

Chipless RFID is, therefore, a technological variant which left behind the communication diagram used by the radio frequency identification systems. In fact, in a chipless RFID system, the tag does not contain an electronic component. No communication protocol is, therefore, possible, unlike conventional RFID technologies. An amplitude, or a phase, modulation using a synchronization based on a clock is, therefore, not possible. A chipless tag should not need any switching system, which allows the commutation between two complex loads at precise moments.

Family	Range	Coding capacity	Access	Positioning	Cost
Optical barcode	Several cm in line-of-sight	>43 bits	Read	Precise	>0.005 euros
Magnetic	Contact	192 bits	Read/Write	Precise	>0.1 euros
Passive UHF RFID	0–7 m	Several kbits	Read/Write	Almost anywhere	>0.1 euros
Passive HF RFID	<1 m	Several kbits	Read/Write	Almost anywhere	>0.4 euros
Printable chipless	<1 m	Several tens of bits	Read	Almost anywhere	>0.005 euros

Table 2.1. *Positioning of chipless technology compared to other identification technologies*

The key element which has initiated the development of this technology is primarily the possibility to reduce tag production costs. This is particularly expected today to the extent that this hinders the large-scale development of RFID systems. In fact, in the price of a conventional RFID tag, there is the cost of the chip that does not cease to decrease, the cost of the antenna and its support, and the cost related to the connection of the antenna with the chip. On the contrary, the price of a chipless tag is only related to the metal deposit and its support. In terms of the production process, we will see below that it is possible to print the tag on its support like a barcode, provided that we use a conductive ink [SHA 11]. Thus, the production cost in this case is almost similar to that of a barcode. Table 2.1 allows us to compare chipless technology with other identification technologies in relation to several essential criteria, such as reading range, storage capacity or even production cost. In terms of the unit cost, a chipless tag can become as competitive as a barcode. It can be remotely read as an RFID tag with a less precise positioning unlike the optical barcode. However, a chipless tag can only be read because it does not contain a memory. In fact, its ID is linked to its geometry. Figure 2.1 allows us to classify different chipless technologies within the RFID scope. Chipless technologies are passive, and can be printable or not depending on the substrate used. Of course, the least expensive tags are those that can be printed, even directly on the object to be traced.

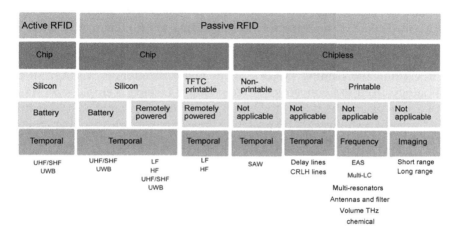

Figure 2.1. *Classification of different RFID technologies according to these criteria (from top to bottom): active/passive, with or without a chip, printable/non-printable, power supply mode and data or ID generation technique*

The physical principles of operation of chipless RFID tags are diverse and can be classified according to the way in which the electromagnetic signature of the tags is used. The approaches used are most of the time temporal, frequential or based on a radar imaging principle. In all cases, the coding of a chipless RFID tag depends directly on its geometry. The latter provides a specific, even unique, signature. We can, therefore, distinguish three chipless tag categories:

– in the case of the temporal approach, the complex load connected to the antenna is usually composed of a transmission line on which parasitic elements are positioned (capacities, etc.) to create reflections at precise moments. The position of each reflector varies from one configuration to another, to achieve a pulse position modulation (PPM) coding;

– the spectral or frequential approach consists of the information coding, by creating peaks or dips related to the presence of a resonance at certain frequencies. They, therefore, require a wide spectrum. The coding used is a transposition of on off keying (OOK) or temporal PPM coding in the frequency domain. Thus, the presence (or absence), or even the resonance shift at a particular frequency constitutes the coding element;

– two-dimensional (2D) imaging can be used when the wavelengths of the frequencies used are small compared to the size of the object to be detected. Its 2D image can be obtained provided that spatial diversity is sufficient. Detection systems for tags operating at 60 GHz can be designed according to this idea.

Regarding the reading system, again major differences distinguish this technology from the conventional technology. In fact, a chipless RFID tag can be seen as a static radar target with a specific electromagnetic signature. The design of a chipless RFID reader is, therefore, very close to that of an air radar, which allows us to detect the signature of flying vehicles, at a scale ratio and of similar power.

A possible alternative that will allow us in a more distant future to decrease the manufacturing costs of an RFID tag is the printed electronics. Tags produced in thin film transistor circuit (TFTC) printed electronics are halfway between conventional RFID and chipless RFID. From a conceptual point of view, there is no major difference with the operating principle used in conventional RFID. However, the manufacturing process using the printing method and the production cost should approach that of chipless RFID.

2.2.1. *Temporal tags*

The first developments which took place in chipless RFID technology were inspired by the operation of conventional RFID tags which code the information as a function of time. Due to the absence of any logical sequencer in a chipless tag, the approach which was undertaken to code information consisted of adapting the type of pulse coding used in certain radio remote controls. For this, it is enough to route the incident wave of the reader toward a delay line on which discontinuities are arranged to reflect a part of the pulse at precise moments. The difficulty of chipless tags based on a temporal coding approach lies in the miniaturization of the delay line in order to code maximum information in a reduced surface.

2.2.1.1. SAW tags

The first chipless tag designs, which were temporally coded, were very logically oriented toward the use of piezoelectric substrates, such as quartz (SiO_2) or lithium niobate $LiNbO_3$. In fact, SAW filters are typically used to create RF filters of reduced size and important order. In a SAW filter, the piezoelectric effect contributes to the transformation of an electromagnetic wave into an acoustic wave. For this, a pair of conductors conveying the electromagnetic wave is connected to a transducer formed by interdigitated electrodes placed on the piezoelectric substrate. The acoustic wave propagation speed is 3,000–4000 m/s [FIN 10], or 100,000 times slower than the speed of light. This makes it the ideal component to create a delay line.

SAW tags are composed of an antenna, often a dipole antenna, directly connected to an electro-acoustic transducer. The operating principle is as follows. A wave in the form of a short electromagnetic pulse is sent by the reader and is captured by the tag antenna. The acoustic wave then propagates at a low speed in the substrate. Reflectors are positioned throughout its path, in order to generate reflections in the direction of the antenna. These reflections will, therefore, be converted into electromagnetic waves and will be reradiated toward the reader. The reflectors are metal elements placed on the piezoelectric substrate and can have different forms and geometric associations.

The design of SAW tags proposed by Nysen et al. [NYS 88] uses several transducers as reflectors, all connected to the antenna (see Figure 2.2(a)). Thus, as soon as the acoustic wave reaches one of the reflectors, it is converted into an electromagnetic wave and returns to the antenna. The modification of the tag ID is performed by changing the spacing between the different transducers. This solution allows us to reduce the insertion losses compared to the second design presented below (see Figure 2.2(b)).

The second principle of SAW tags (which is still the most commonly used) is based on a single transducer and several reflectors in open circuit (see Figure 2.2(b)). Unlike multi-transducer tags, the reflected waves must travel in the opposite direction to then be retransmitted by the antenna. It is especially on this principle that SAW tags operate, marketed by RFSAW [HAR 02]. The coding used is based on a pulse-position modulation combined with phase information which allows us, among others, to increase coding effectiveness on an effective area. Coding is, therefore, performed by modifying the positioning of the reflectors on the substrate. This system proves to be efficient: a coding capacity of 256 bits is possible and its dimensions can be relatively reduced (1×10 mm², excl. antenna). The antenna used is typically a dipole or a monopole, which provides a length close to 30 mm at 2.45 GHz for the latter.

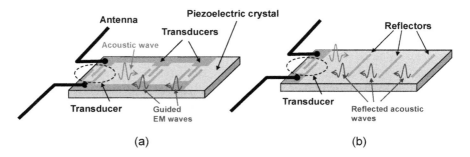

Figure 2.2. *a) SAW tag proposed by Nysen et al. [NYS 88]; b) RFSAW tag developed by Clinton et al. [HAR 02]. For a color version of this figure, see www.iste.co.uk/vena/chipless.zip*

Finally, the features of sensors are possible with a SAW tag [FIN 10]. For this, it is enough to add a transducer directly connected to a sensor with a resistive or a capacitive load, which is sensitive to an environmental parameter. It is thus possible to detect the temperature with a thermistor. Another process uses a resonator whose frequency is a function of the ambient temperature.

Despite their interesting performance, today SAW tags are not competitive in terms of cost, with a unit cost of 10–20 cents. However, a quite recent patent studies the possibility of replacing the piezoelectric substrate with a specific cellulosic substrate [KIM 08], where the arrangement of the fibers has been modified in such a way that acoustic waves can be created. In addition, the use of SAW tags may prove to be difficult depending on the operating environment. In fact, piezoelectric substrates are very susceptible to electrostatic discharges.

2.2.1.2. *Tags using transmission line*

Other solutions, which are still at the research stage, are proposed in the temporal field. The reading principle remains the same in the sense that the reader sends an ultrashort wave and the tag responds by multiple reflections as a radar echo. The substrates used are more conventional and their cost is in particular potentially low.

In 2006, Zhang *et al.* [ZHA 06] proposed a delay line on which discontinuities can be induced by located or distributed elements, in order to create reflections (see Figure 2.3). The delay line is of a microstrip type and forms meanders in such a way so as to maximally reduce the total tag surface. An antenna must be connected to this delay line to create the tag. Each reflection occupies a slot of 2 ns, therefore, a minimum spacing of 180 mm between each discontinuity is necessary to prevent pulse overlapping. Discontinuities are created by localized capacitive elements. Based on their presence, a reflection will take place or not. Thus, in a surface of 8.2 × 3.1 cm², there are four delay line sections. This represents a coding capacity of 4 bits. The transmission pulse used for the measurement has a width of 0.5 ns.

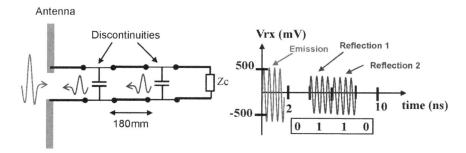

Figure 2.3. *Operating principle of the tag proposed by Zeng et al. [ZHA 06]. For a color version of this figure, see www.iste.co.uk/vena/chipless.zip*

In 2008, Zheng *et al.* [ZHE 08] developed the design by replacing the localized elements which allow the creation of discontinuities by distributed elements, making the structure compatible with a production process by the printing method. The authors presented a technical method to configure tags by adding a localized material with the use of a conductive ink inkjet printer. A coding capacity of 8 bits is achieved, which is 2 times higher than that announced in [ZHA 06], but, on the other hand, the necessary surface is also doubled. This design is still too bulky to be used in the framework of a traceability application for consumer goods.

By maintaining this principle, in 2009, new studies [MAN 09, SCH 09] presented a solution to reduce the length of the delay line sections. To slow the wave propagation, left handed (LH) transmission lines were used. However, LH transmission lines cannot be produced in a simple manner. In fact, compared to the electric model of a transmission line, in an LH line, serial and parallel components are reversed. The tag presented in [MAN 09] and [SCH 09] uses capacitors in series and inductors in parallel connected to the ground plane. The decrease in the line length is, therefore, possible by increasing its implementation complexity, which increases the final cost of the tag implementation. An interesting design introduced in these articles concerns the coding principle which is used based on a phase modulation in a temporal manner. In the end, a coding capacity of 6.1 bits is achieved for a structure 26 cm long. The specific processes described in [SCH 09] incorporate the same line structure by adding a discontinuity that can act as a sensor if a capacitive or inductive component which is sensitive to an environmental parameter is used as a Barium Strontium Titanate (BST) capacitor sensitive to temperature.

In 2006, Chamarti et al. [CHA 06] proposed a delay line that allows us to create an ID for chipless RFID sensors and tags. The notable difference compared to previous designs comes from the fact that the line is used in transmission. Two ports are used, one on each end of the line. The incoming signal is divided into two parts between a delay line and a non-delayed direct line. The direct line routes the input pulse without delay. An isolator connects the delayed line at the non-delayed line to direct a second pulse at a precise moment. Depending on the isolator position, the position of the second pulse also varies, which constitutes a coding technique in pulse-position. In 2007, this concept was applied for the implementation of a chipless tag operating at a frequency of 915 MHz [VEM 07] (see Figure 2.4(a)). A circulator connected on one side to the patch antenna and on the other side to the in/out ports of the filter is used to isolate the transmission/reception signals. A coding capacity of 2 bits is obtained, which allows us to demonstrate the proof of concept. Finally, in 2009, the authors proposed a more conventional design [SHR 09], shown in Figure 2.5(b), using a delay line (of meander type) in reflection as for [ZHA 06, ZHE 08, MAN 09, SCH 09]. Coding is still, however, based on a pulse-position approach. In addition, an ethylene sensor feature is added to the tag, by connecting at the end of the line a capacitor sensitive to the environment. The modification of its value introduced a phase change at the response level of the tag antenna mode, with a sensitivity of 26.5°/pf.

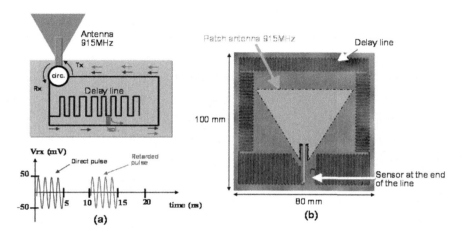

Figure 2.4. *a) Operating principle of the tag introduced by Vemagiri et al. [VEM 07]; b) picture of a chipless tag by Shrestha et al. [SHR 09] incorporating a sensor at the level of its termination. For a color version of this figure, see www.iste.co.uk/vena/chipless.zip*

These preliminary studies allowing primarily to provide a proof of concept have been very widely completed thereafter, in such a way so as to show the potential of this approach [GIR 12]. On this subject, we highlight some developments covering, of course, also the theoretical aspects to retrieve the tag information in a real environment [LAZ 11], as well as some processes at the reader side, to show the potential application of this approach [RAM 12].

A study on C-sections, which were used to obtain dispersive devices where frequency evolution of *group delay* could be imposed [GUP 10], has allowed us to consider the use of these structures for chipless applications. Thus, a chipless "circuit" approach, where the tag is composed of two antennas each connected to a C-section, has allowed us to develop a new family of chipless tags [NAI 13a]. The tag produced is based on the operating principle of "C-sections" which can be seen as extremely compact delay lines and whose dispersive nature will allow the increase in the coding capacity. We can observe a C-section line in Figure 2.5(a). The latter contains a group of 3 "C-sections"cascaded with a "group of 10 C-sections". In addition to the structure in "C-section", the chipless tag consists of two wideband antennas: both can be used (according to the tag orientation compared to the reader antenna) for transmission and/or reception purposes. As shown in Figure 2.5(b), both antennas are positioned perpendicularly to one another in such a way as to ensure a better isolation of the return signal.

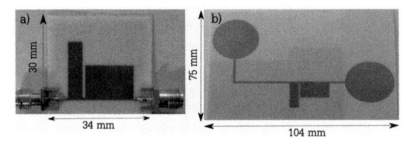

Figure 2.5. *Example of a multi-group structure of "C-sections", i.e. of a line consisting of several groups (two in this case) of, respectively, three and 10 "C-sections". a) Configuration with SMA ports and b) multi-group chipless tag*

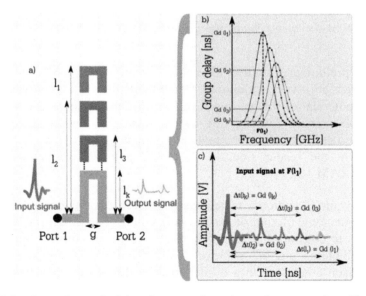

Figure 2.6. *Operating principle of a tag based on the use of a "C-section": a) diagram of a C-section, with the length li being variable; b) group delay as a function of frequency for different lengths of li line; c) correspondence in the temporal field. For a color version of this figure, see www.iste.co.uk/vena/chipless.zip*

A "C-section" of l_1 length produces a maximum delay for each multiple-length odd-number of λg / 4, where λg is the guided wavelength (see Figures 2.6(a) and (b)). We can approximate the frequency for which this maximum value is produced using the expression: $F(l_1) \approx c/(4* l_1 (\varepsilon_{reff})1/2)$ where ε_{reff} is the effective permittivity of the microstrip line, and c is the speed of light. Any other line length l_i, $i = 2, 3,\ldots, n$ will generate a different delay, considerably lower than l_1 at the $F(l_1)$ frequency (see Figure 2.6(c)). From that point, when the wave transmitted by the reader arrives at the tag, a part is directly reflected by the object (we will discuss the structural mode), whereas another part will be captured by the antenna and propagated along the "C-section" to be finally reradiated by the second antenna. To the extent that the duration of the wave propagation through the "C-section" is a function of the length l_1, the reader will recover two signals (the structural mode and the antenna mode) shifted in time. This time difference will allow us to code the tag ID number. Thus, if we consider n tags containing, respectively, $l_1, l_2 \ldots Ln_n$ lengths, each will produce different IDs, which can be noted asID1 = Δt (l_1), $\ldots$, IDn = Δt (l_n), where Δt (l_i) is the value of the temporal shift between the structural mode and the antenna mode of the ith configuration. We observe here that this device belongs to the family of temporal tags. The information is directly visible on the temporal signal reflected by the tag. In this configuration (a single group of "C-section"), the main interest lies in the miniaturization of the device compared to a conventional line length or even to a meander line. In [NAI 13a], a study on this point shows that for a *group delay* value of 1 ns, the device at "C-section" has a surface 6 times lower than that in meanders. In addition, if the line is expanded, its total length is almost 2 times shorter than the meander line or even the straight line. We also show that for the dimensions of 15 × 30 mm², it is possible to obtain very considerable delays, namely, of 10 ns. It is also possible to cascade several groups of "C-section", so as to increase the quantity of information present on a tag [NAI 13b].

2.2.2. Frequency tags

The second approach used is based on the frequency signature of the tag for the information coding. Unlike a temporal system, the information contained in the tags is based on a variation of amplitude or phase as a function of frequency and not anymore as a function of time. This approach seems to be the most promising in terms of the quantity of information that it is possible to implement in a chipless tag. The surface information density is also more appropriate than for the temporal approach. However, the bandwidth required to code a large number of bits is problematic in relation to transmission standards imposed by the Federal Communications Commission (FCC) in the United States and the Electronic Communications Committee (ECC) in Europe. Different techniques have been explored to obtain spectral signatures of chipless tags.

1-bit tags used in the anti-theft systems Electronic Article Surveillance operating at HF, UHF and SHF frequencies are surely the precursors of chipless tags in spectral signature. It is only very recently (since 2002 [FLE 02]) that the first frequency chipless tags offering a coding capacity greater than 1 bit have been proposed. Since then, different variants have been introduced. We can classify them as follows:

– First, the frequency tags in a passive circuit are composed of a ultra-wideband signal (UWB) receiving antenna and a UWB transmitting antenna which are connected to a filter circuit. The filter created with the use of distributed elements acts as a multi-band notch filter. The tag identification is in this case related to the spectral response of the filter used.

– A variant to the previously mentioned concept uses one or several antenna(s), which operate in both transmission and reception, and which are connected to a complex load. In this case, the modification of the complex load will allow us to change the tag ID.

– Then came the frequency tags based on the association of multiple filter antennas which act as the receiving antenna, the transmitting antenna and the filter circuit. This approach is thoroughly discussed in this book.

– Finally, we can find a last variant of frequency tags in THz frequency ranges, whose specificity is to code the ID no longer on the surface with 2D metal arrangements, but in the tag volume, particularly with multilayer structures.

2.2.2.1. *1-bit tags for anti-theft systems*

1-bit tags used in the EAS systems constitute the chipless technology which has experienced the first commercial success globally. The developments that take place today around chipless RFID technology can be considered as a generalization of the concept for applications that no longer require "theft/anti-theft" binary information, but a unique ID or alias name in tens of bits. It is a frequency coding that is used for the 1-bit tags, and the operating frequencies may vary from the HF band to the microwave frequencies.

In HF, in the same way as for a conventional RFID system, the interaction mode between the detector and the tag is based on an inductive coupling [FIN 10, FLE 02]. The tag is composed of a loop antenna connected in parallel with a capacitor, so that together they constitute a resonant circuit at a very precise frequency [2.1] as presented in Figure 2.7(a).

$$f_r = \frac{1}{2\pi\sqrt{LC}}$$

[2.1]

According to the quality factor of the resonant circuit, a more or less pronounced frequency selectivity will be obtained. To detect this type of tag, the reader must perform a scan around the resonance frequency. A sensitive receiver located at the detector level will pick up the signal which is retransmitted by the loop as a function of frequency. If the tag is present, a resonance peak followed by a resonance dip is detected [FIN 10]. TagSense proposes an anti-theft detection system based on this principle. LC-10 [TAG 16] reader performs a scan at a frequency between 1 and 50 MHz. It can potentially detect up to 64 resonances in this frequency band, which can be from multiple 1-bit tags or from a multi-bit tag with several resonators.

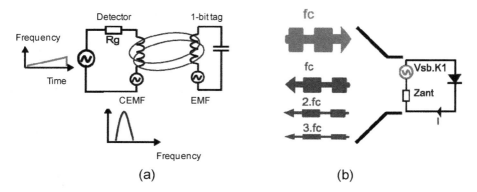

(a)   (b)

Figure 2.7. *Operating principle of a 1-bit tag: a) operating in HF frequencies and b) operating in UHF / SHF frequencies. For a color version of this figure, see www.iste.co.uk/vena/chipless.zip*

Other physical principles have been used for the design of 1-bit tags, in particular for the microwave frequencies, where in this case, the interaction mode between the tag and the detector is based on the backscattering of the propagated wave which is transmitted by the reader as in the case of UHF/ SHF RFID. To separate the signal of the detector from that of the tag, the solution used is to generate, at the tag level, harmonics of the carrier wave frequency, by using a non-linear component such as a diode [FIN 10] (see Figure 2.7(b)). The frequency commonly used is 2.45 GHz. In this case, a signal can be retrieved at the second and the third harmonic, i.e. at 4.9 and 7.35 GHz if the 1-bit tag is close to the detector. In fact, in reception, we can model the tag antenna by a radiation resistance in series with a monochromatic voltage source [2.2] whose amplitude depends on the power level that is picked up by the antenna (see Figure 2.7(b)). This generator is connected to a diode whose characteristic voltage current I(V) can be approached by a third-degree polynomial [2.3]. The current passing through the radiation resistance can be expressed as a function of the generator voltage as follows [2.4]:

$$V(t) = U \cdot \cos(\omega t) \qquad [2.2]$$

$$I(V) = a \cdot V^3 + b \cdot V^2 + c \cdot V \qquad [2.3]$$

$$I(t) = \frac{b \cdot U^2}{2} + \left[c \cdot U + \frac{3.a \cdot U^2}{4} \right] \cdot \cos(\omega t) + \frac{b \cdot U^2}{2} \cdot \cos(2\omega t) + \frac{a \cdot U^3}{4} \cdot \cos(3\omega t) \qquad [2.4]$$

Therefore, we can see that in equation [2.4], a signal appears at 2 and 3 times the frequency transmitted by the reader. The a and b terms of the polynomial depend on the type of diode used, and according to their values, the harmonics will be more or less pronounced. Due to its compactness, this detection principle can be adapted to detect avalanche victims (RECCO system) or to track animals, and even insects [KIR 07].

2.2.2.2. *Tags using two antennas and a passive filter circuit*

This chipless tag design was introduced by Preradovic *et al.* in 2008 [PRE 08]. The basic idea is to use a receiving antenna which will pick up a UWB signal. This signal is then transmitted to a configurable band notch filter. The filter output is finally connected to a transmitting antenna, so that the filtered signal can be returned by the tag to the reader. Depending on the chosen configuration, the filter will allow us to generate a unique "electromagnetic" signature. In order to reduce the coupling between the transmitting and receiving antenna at the reader and the tag level, these two are positioned in cross polarization, as we can see in Figure 2.8(a). Thus, the transmission signal is vertically polarized and the reception signal is horizontally polarized. In addition, this allows us to reduce the interference related to reflections of the detection environment. The tag antennas are UWB monopoles which are designed to operate for frequencies ranging from 2 to 2.5 GHz. The multi-band notch-filter is created with the use of distributed elements, by using a microstrip [PRE 08] or a coplanar [PRE 09a] transmission line. To reject certain frequencies, spiral resonators are placed on either side of the transmission line. These resonators act as short-circuits to ground at their resonance frequencies. The identification of a particular tag is a function of the presence of a rejection at a particular frequency. To vary the spectral signature of a tag to another, it is, therefore, sufficient to add or remove resonators. The solution used in this design is to remove the resonators by placing metallic short-circuits on the spirals, in order to push their resonance frequencies outside the bandwidth of the detection system, as shown in Figure 2.8(b). Figures 2.8(c) and (d) represent the amplitude and the phase in the filter output for two different configurations. S21 represents a ratio of a complex power root between the power transmitted by the source and the power measured in reception. This parameter can be measured by the vector network analyzer which is the basic element of the frequency bistatic radar testbed described in Chapter 5. Thus, in the case of a filter, the absolute value of S21 parameter represents directly the insertion losses. The coding used in this case simply allows us to associate a bit with a resonator, which provides a coding capacity of 6 bits in this specific case. Thus, when the resonator does not have a metallic short-circuit, a dip is visible in the spectrum, which corresponds to the binary value "1". However, when a metallic element shunts a part of the resonator, the dip is no longer visible, which corresponds to the binary value "0".

The proposed reading system [PRE 09b, PRE 10a] to detect these tags incorporates the operating principle of a vector network analyzer, detailed in Chapter 5. A non-modulated sine wave carrier whose frequency varies by constant steps is sent by a transmitting antenna in vertical polarization. The tag picks up this signal and reflects a part with a level that depends on the configuration of the multi-band notch-filter. This signal fraction is transmitted in cross polarization and is picked up by the receiving antenna of the detection system. A power and phase detector is used in order to obtain the amplitude and phase characteristics as a function of frequency. We will see below that this detection approach leads to the transmission of CW signals that can hardly conform to the transmission standards.

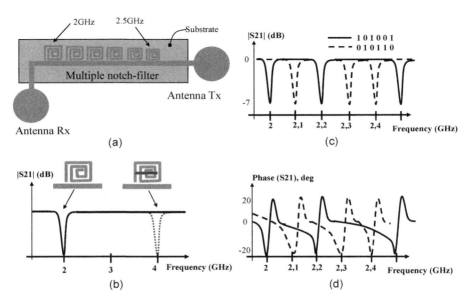

Figure 2.8. *Operating principle of a tag with double antennas in cross polarization introduced by Preradovic et al. [PRE 08]; a) picture of a 6-bit tag. A ground plane is present under the substrate; b) configuration method of spiral resonators with the use of a short-circuit; c) insertion losses of a multi-band notch-filter for several configurations; d) phase behavior of the filter*

Subsequently, new structures, which maintain the same operating principle, have been proposed. Thus, a structure of 35 resonators coding 35 bits in a surface of 88×65 mm^2 has been obtained [PRE 09c]. This coding capacity is close to the capacity that we can obtain with an optical barcode. Remember that the EAN 13 barcode has a coding capacity of 43 bits. In order to reduce the tag surface that increases proportionally with the number of resonators, the transmission line is no longer in a straight line but meanders. In addition, spiral resonators are distributed on both sides of the transmission line. The filter bandwidth necessary to code 35 bits is between 3 and 7 GHz. This allows us to be compatible with the FCC regulation which authorizes that UWB communications are spread out between 3.1 and 10.6 GHz. The tag, however, remains bulky and is not compatible with a practical use, in particular because of the high sensitivity of the antennas when the tag is positioned on an object.

The different designs mentioned so far have a ground plane. In order to make this design potentially printable directly on the objects to identify, a variant using the *co-planar waveguide* (CPW) coplanar transmission lines has been proposed [PRE 09a]. Thus, a capacity tag of 23 bits on a surface of 108×64 mm² (incl. antennas) has been designed. The tag has been produced on a flexible substrate 90 μm thick. The filter bandwidth is spread out between 5 and 11 GHz. Starting at 5 GHz, instead of 3 GHz as in the past, allows us to compensate for the increase in the surface required for the spiral resonators, which can no longer be distributed on either side of the line as in the case of a microstrip line.

The latest generation of this type of tags allows us to use a single transmitting / receiving antenna [PRE 10b], in particular to make the system less sensitive to the tag orientation, as well as to reduce its surface. These studies remain, however, exploratory, and to this day, only a capacity of 6 bits has been achieved.

Other studies incorporating this basic principle have been made, particularly in order to miniaturize the multi-band notch-filter. Thus, in [BAL 09a], we see a tag with a UWB receiving antenna, in a monopole form which is connected to multiple folded narrowband dipole antennas, of different resonance frequencies (see Figure 2.9).

These dipole antennas act both as transmitting antennas and as filter antennas. To modify the spectral signature of the tag, it is sufficient to remove one or several dipole antenna(s). A coding capacity of 6 bits, for a surface of 40×40 mm² is achieved. The structure requires two structured copper sides, which makes it not very competitive in terms of production cost.

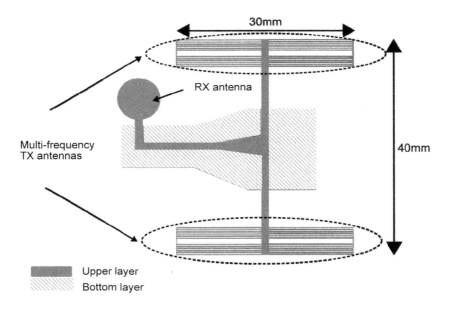

Figure 2.9. *Chipless frequency tag developed by Balbin et al. [BAL 09a].*
For a color version of this figure, see www.iste.co.uk/vena/chipless.zip

2.2.2.3. *Tags using one or multiple antennas connected to a complex load*

The separation of the receiving and the transmitting antenna at the tag level has the advantage of limiting the interference between the received signal and the electromagnetic response of the tag, especially by using two orthogonal polarizations. However, in some cases, this solution remains costly in terms of surface and the number of maximum resonators is limited by the insertion losses of the filter. Other designs have been proposed in which one, or even several, antennas are used in both reception and transmission. Again the antennas do not participate in the information coding function, but it is the complex load connected to their terminals which will allow us to differentiate the spectral signature of the tag.

Thus, in 2007, Mukherjee [MUK 07a, MUK 07b] introduced a chipless tag design composed of an antenna and a parameterized complex load (see Figure 2.10(a)) allowing us to modify the value of the reflected signal phase as a function of frequency, as presented in Figure 2.10(b).

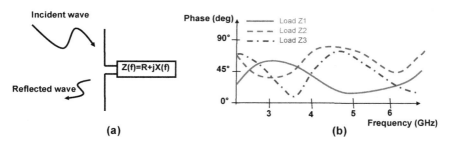

Figure 2.10. *Design of an antenna connected to a complex load introduced by Mukherjee [MUK 07a, MUK 07b]*

In the proposed solution, the frequency band used is between 6.9 and 7.9 GHz which implies the use of a UWB antenna. The proposed complex load is a distributed LC filter, which is implemented with a microstrip line. The load size is 5 × 12 mm², to which we must add the UWB antenna surface (20 × 20 mm²), which gives a relatively reduced size compared to previous designs. However, the coding capacity remains very limited.

Another tag approach is also based on the phase variation for the information coding [BAL 09b]. Unlike Mukherjee who uses a single UWB antenna connected to a complex load, the authors use multiple narrowband antennas connected to variable-length stubs in open circuit. For each antenna, as a function of the stub length, a relative phase variation, compared to cases where no stub is connected, is visible at the resonance frequency of the antenna. To validate this design, three patch antennas at 2.15, 2.3 and 2.45 GHz frequency have been used, as we can see in Figure 2.11.

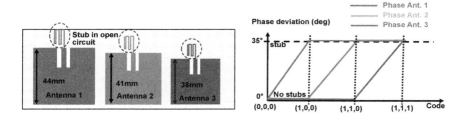

Figure 2.11. *Design of narrowband multi-antenna tag connected to variable-length stubs. ID generation depends on the phase. For a color version of this figure, see www.iste.co.uk/vena/chipless.zip*

To detect a relative phase variation, a reference measurement is necessary. Thus, the measurement of the tag electromagnetic response in horizontal polarization will allow us to obtain this phase reference. The tag response phase in vertical polarization depends on its loads or stubs, which are connected to the antennas. It is, therefore, the phase difference between vertical and horizontal polarization which will be used to trace the tag ID (see Figure 2.11). This detection principle seems to bring a certain reading robustness, since a differential measurement taking advantage of the polarization diversity is performed. However, the size of this tag is still large in relation to its coding capacity (3 bits). It is also necessary to perform two measurements, a procedure which complicates the reader part.

2.2.2.4. UWB tags to multiple filter antennas

The ultimate step to minimize the chipless tag surface is to assemble the receiving antenna, the transmitting antenna and the filter in a single element, the coding RF particle (or even resonant diffractive element) on the basis of its own RCS.

In 2005, Jalaly *et al.* [JAL 05a] proposed a very simple solution, involving multiple dipoles loaded by variable-value capacitors, in order to create multiple resonances in the spectrum. A variant to the first proposed solution used dipole resonators with variable-length ground plane, in short-circuit [JAL 05b], as shown in Figure 2.12(a). This variant is more interesting because it allows a low-cost manufacturing process with a conductive ink inkjet printer. This tag topology closely resembles optical barcodes by their form, but also the coding approach which consists of the modification of vertical bar geometry for the information coding. For this reason, the authors have introduced the term "RF barcode" to describe this design and generally this chipless RFID technology. As described in [PER 14], this tag design can be considered as the origin of *RF encoding particle* chipless tags. These are tags which currently have the best performance in terms of coding density and size. These tags are at the heart of this book and will be discussed in Chapter 4.

The dipoles composing the "RF barcode" introduced by Jalaly *et al.* [JAL 05b] are, however, placed on a substrate with a ground plane. They behave as short-circuited "patch" antennas, which resonate at a half-wavelength. A model of this resonator type can be obtained by considering a microstrip line in open circuit at its two ends. It is precisely at the level of these apertures that reradiation will take place. Each aperture can be modeled by a slot antenna [BAL 05]. As it is shown by the electromagnetic signature received by the detection system, Figure 2.12(a), at each resonance, we can observe a resonance dip and not a peak as we could imagine. In fact, this dip is the result of a destructive interference between the response of the single ground plane and that of the dipole at the resonance [VEN 13, KAR 10]. This phenomenon is, therefore, visible with other chipless tag designs with a ground plane. This point will be analyzed in detail Chapter 4.

The coding capacity reached by the structure proposed by Jalaly [JAL 05b] is 5 bits, by using five resonators for a relatively reduced surface of 25 × 30 mm². The necessary bandwidth is spread out between 5.45 and 5.85 GHz, or 1 bit for 100 MHz. Measurements show that it is possible to detect this tag at a distance of up to 2 m, for a transmitting power of 500 mW.

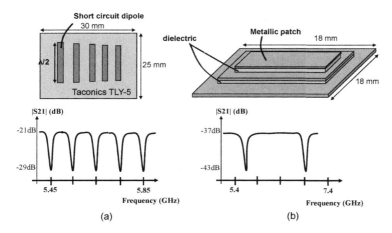

Figure 2.12. a) "RF barcode", or even dipoles in short-circuit with ground plane [JAL 05b]. b) Dual-band multilayer patch [MUK 09]

In the same idea, Mukherjee in 2009 transposed the design of resonant patch antennas by Jalaly in a three-dimensional form [MUK 09]. Rather than positioning the dipoles side-by-side above a ground plane, the idea was to superimpose multiple "patch" antennas (see Figure 2.12(b)). Thus, according to the frequency, a metal rectangle can resonate or act as a ground plane for a resonator located above him. Thus, a patch consisting of three layers can resonate at two different frequencies. The coding capacity aspect is not addressed in the paper, but the possibility of transforming this tag into a temperature sensor for severe environment conditions is discussed. The tag surface is 18×18 mm^2 for usage frequency between 5.4 and 7.4 GHz.

We thus observe that the previous designs require at least two conductive layers to generate an electromagnetic signature with relatively selective resonance peaks. However, there is an interest in not to including a ground plane to simplify the tag manufacturing process. Thus, in an article dating back to 2006, McVay *et al.* [MCV 06] presented a chipless RFID tag, which did not include a ground plane, based on the association of five resonators, as we can see in Figure 2.12(a). The frequency band used is spread out between 0.7 and 0.9 GHz. In order to decrease the size of resonators which become consistent with these frequencies, Peano and Hilbert fractal curves are used, respectively, at the second and third order. Thus, the largest dimension of resonators is 0.07 λ. The occupied surface that contains the five resonators is 150×30 mm^2. A study on the tag sensitivity in relation to the support on which it is positioned is performed in this article. The authors show that a tag without a ground plane can be used to identify a disruptive object as a roll of paper ($\varepsilon_r = 2.6$, tan δ = 0.08).

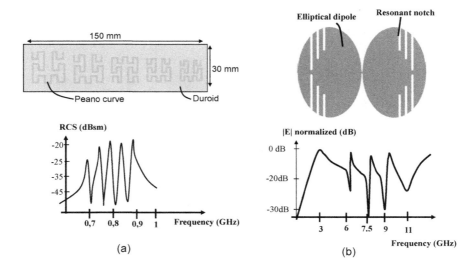

Figure 2.13. *a) Multi-resonator tag based on Peano curves at the second order [MCV 06]; b) elliptical notch dipole by Blischak [BLI 09]*

Another design without a ground plane and operating in the UWB frequency range was proposed in 2009 [BLI 09]. The presented structure is an elliptical dipole with a signature in the UWB band (see Figure 2.13(b)). Notches are performed in the dipole arms, in order to create notch-filters of integrated bands at the antenna. The electromagnetic signature of the tag presents dips at certain frequencies that depend on the notch length. In the example studied, three notches are used, which gives us a coding capacity of 3 bits. A tag electromagnetic signature modeling in the form of poles and residues is presented. The basic hypothesis is based on the fact that any impulse response $h(t)$ can be modeled by a dumped exponential sum characterized by poles and residues [2.5]. Each pole of k grade contains information on ω_k pulsation and m_k damping, while the residue contains information on A_k amplitude and Φ_k phase.

$$h(t) = \sum_{k=1}^{p} A_k \cdot e^{j\Phi_k} \cdot e^{-m_k t} \cdot e^{j\omega_k t} \qquad [2.5]$$

To determine the poles and the residues associated with a tag, a pole extraction technique called "Matrix Pencil" is applied on the tag impulse response. The interesting aspect of this model is that pole damping and pulsation are independent from the observation point and from the level of the electromagnetic field transmitted by the reader [BLI 09, LEE 05, REZ 14], unlike at the amplitude and phase contained in the residue. This technique can, therefore, be useful to extract tag-specific information in a detection environment which is unknown due to several variables (reading distance, tag positioning, etc.). However, it is challenging to design resonators to make a chipless tag, where we can control the damping coefficient for the information coding. In addition, the method requires a certain computing power to extract tag information, which in practice will create constraints at the reader side, in particular regarding reading time which must be almost instantaneous as for all competing identification techniques.

Finally, in 2010, a chipless tag based on split ring resonators (SRR) was introduced [JAN 10]. The chipless tag is thus seen as a multi-band frequency selective surface. A network of 10 SRR of identical size is used to reject up to four frequencies. Depending on the gap orientation of different SRR, rejection frequencies vary, and we can observe up to four dips in the frequency spectrum. The authors, therefore, estimate a coding capacity equivalent of 4 bits contained in a surface of 10×20 mm. The SRR network has been produced by the printing method on a plastic support in the form of a credit card (85×53 mm^2). To read the tag ID, a reading system, of an 8–12 GHz frequency, based on the use of an open waveguide, has been tested. The interesting aspect of this technique is that the transmission signal is confined in the waveguide, which allows us to respect the regulations imposed by the FCC or the ECC. However, the reading of the tag ID is performed by contact. This principle, which is based on the use of a waveguide, can, therefore, be used for person identification applications (ticket, etc.), but it will be difficult to transpose it for object identification, where a reading range in meters is preferable.

However, this concept of a chipless tag seen as an FSS has also been adopted more recently [COS 13, COS 14] for reading applications in the far-field. In this case, the tag is simply a necessarily truncated periodic surface, which has the behavior of a filter that is apparent at the RCS level of this structure of frequency dips, which will allow the information coding [COS 13]. Each periodic element is composed of resonators embedded in one another as in [VEN 11], which allows us to miniaturize the whole system size. The periodicity will allow us to simplify the design phase to the extent that the unit cell can only be optimized, by considering the periodic walls on the sides. Similarly, in a certain way, this periodicity will also allow us to increase the tag RCS and, therefore, facilitate its reading.

2.2.2.5. Volume-coded THz tags

Most of the proposed chipless tags are in UHF (0.3–3 GHz) and SHF (3–30 GHz) frequency ranges. The coding involved is primarily based on resonance phenomena. The size of the conductor elements is of the order of a half-wavelength. By increasing the working frequency, the tag dimensions will inevitably become smaller. The other benefit of increasing the frequency is to be able to avoid RF transmission standards, in particular in THz frequency ranges (0.1–100 THz). Thus, the first research works concerning the development of a chipless RFID tag in THz frequency ranges emerged in 2010 [TED 10, BER 11, PER 11]. This chipless tag operates according to the photonic crystals principle, which allows the rejection of frequency bands (band gaps), in the same way as notch-filters of the bands set out above. The structure proposed in Figure 2.14(a) is composed of alternating material layers with different optical grades, which is a Bragg reflector. The center frequency and the band gap width depend on the width of the layers, the material grades and the number of layers. By working at large frequencies, as in this case with several hundreds of GHz, this layer stack will have a total thickness of a millimeter, which is compatible with a chipless use. Thus, the packaging of the object to be identified, in this case, can itself be the label containing its ID. We also note that as the first chipless tags did not contain any conductive material, this allowed us to reduce the cost and provide a completely recyclable solution.

In order to differentiate the electromagnetic signatures, the technique used is to create one or more defaults in the periodic structure (see Figure 2.14(b)). They will create one or several peaks in the band gap. In fact, the emergence of these defaults requires breaking the periodicity of Bragg reflector, by modifying the central air layer, or by replacing it with another material. In the example studied in [TED 10], 11 layers are used. The materials are of silicon with an optical index of 3.415, and of air, with an optical grade equal to 1. Their respective thickness is 75 and 255 μm. The first band gap measured spreads out between 150 and 350 GHz. The measurement results observed for four different default configurations show the potential of this technology. The coding capacity announced is 15 bits.

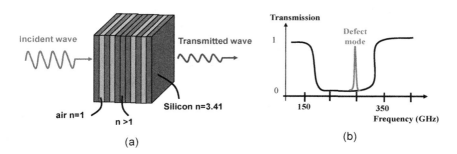

Figure 2.14. *a) Periodic structure based on alternating layers of silicon and air, creating a band gap in THz frequencies [TED 10, BER 11]. b) Transmission characteristic as a function of frequency with a peak in the band gap*

Thus, the first devices were produced by stacking spaced silicon wafers of air gaps [BER 11a]. After that, structures created by paper materials and polymers were obtained [HAM 12, HAM 11, PER 13]. High-grade layers were produced by adding loads (TiO_2, for example) which are commonly used in the paper industry. In summary, the main motivations of this study in the THz domain are (1) the increase in the security of data contained in the tag, (2) the low cost of the solution (without the use of a conductive ink, with processes and materials which are standard in the paper field) and (3) the increase in the coding capacity to the extent that a more conventional chipless RF coding (deposit of a conductive pattern on the stack of layers related to the THz part) can also be added.

2.2.3. *Tags in 2D imaging*

Tags based on temporal or frequential approaches use detection systems which do not require spatial diversity. Measurements can most of the time be carried out in monostatic configuration. In order to increase the amount of information, some studies have sought to transpose radar imaging techniques for the detection of chipless RFID tags. Thus, with this technique, it is possible to trace the geometry of the tags and subsequently its identifier. In 2007, InkSure Technologies [INK 16] developed a chipless tag (SARcode) [HAR 09] that can be detected up to a distance of 2.5 m. The reading system functions as a synthetic aperture radar at 60 GHz to trace the geometry of the metal elements placed on the tag. For this, an antenna array positioned in a semi-circle is used. The tag is composed of small metal elements printed next to one another (see Figure 2.15). The tag ID modification is performed by varying the vertical position of each metal element. The printed patterns are very close and generate diffraction spots which overlap. An image processing software allows us to separate them based on the fact that the spreading phenomena which are related to the diffraction are predictable. Thus, a coding capacity of 96 bits is reached with a surface of 108×15 mm². The patterns used are bars (or short-circuited dipoles) oriented at 45°, so as to transmit an electromagnetic field polarized perpendicularly to the incident field. As the tag can be printed directly on the product and since the coding capacity is relatively large, this solution is promising. However, the tag must be precisely positioned in front of the antenna, so that the diffraction patterns can be analyzed. In addition, the cost of a reading system of this type is relatively high (several tens of thousands of euros) compared to chipless RFID systems operating in RF (< 4.000 euros).

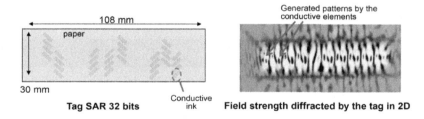

Figure 2.15. *SARCode system developed by Inksure [INK 16, HAR 09]*

2.2.4. *TFTC tags*

Printed electronics attracts significant interest from manufacturers because it is an alternative to RFID and could potentially lower the production costs of conventional RFID tags significantly. The idea in this case is to transpose the equivalent of a chip in a conventional tag in printed TFTC electronic format [SUB 05], but with much lower performance. For example, currently, it is difficult or even impossible to be able to detect TFTC tags with the same readers and the same protocols used for the conventional RFID technology. In fact, the produced "chip" is rudimentary, as it does not have a microcontroller or a reprogrammable memory. The memory size is a few bits. The transistors are printed directly on the tag support. Since 2007, companies such as PolyIC and Kovio [HAR 09] have proposed completely printed RFID tags, which operate in the HF range at 13.56 MHz.

The PolyID tag, which is marketed by PolyIC, has a storage capacity of a few bits. In addition, this memory space does not have a write access. The operating frequencies that can be used with organic printed electronics are limited by the mobility of the charge carriers in plastic conductors (approximately 0.3 cm²/Vs). As a comparison, InGaAs semiconductors are characterized by a mobility of 5,000 cm²/Vs. Consequently, only RFID tags operating at 13.56 MHz are currently produced with organic materials. To overcome the limitations of organic materials, Kovio is developing a printing process involving an inorganic ink based on nano-silicon. Thus, the mobility of charge carriers achieved is 80 cm²/Vs, with a target of 250 cm²/Vs. This should allow the printing of UHF tags in the future. Currently, many challenges still remain, such as the improvement of the ink conductivity and the decrease in their cost, as well as the miniaturization of the tag surface. It, therefore, seems that chipless RFID, which does not use TFTC is currently the solution we have to adopt for low-cost identification applications.

Technology	Memory	Surface	Bandwidth	Manufacturing	Range	Anti-collision
SAW [HAR 02, NYS 88]	256 bits	10 × 10 mm² + ant.	2.45 GHz	Non-printable	30 m	Yes
Delay line [ZHA 06, ZHE 08]	8 bits	82 × 106 mm²+ ant.	UWB	Printable	/	No
Delay line CRLH [MAN 09, SCH 09]	6.1 bits	260 × 30 mm²+ ant.	UWB	Non-printable	/	No
Delay line E/S [CHA 06,VEM 07]	2 bits	NA	UWB	Non-printable	/	No
LC EAS Tag [TAG 16]	1 bit	NA	1–50 MHz	Printable	1 m	Spectral
Tag with double antennas [PRE 08, PRE 09a, PRE 09c]	35 bits	88 × 65 mm²	UWB	Printable	1 m	No
Folded dipole tag [BAL 09a]	6 bits	40 × 40 mm²	UWB	Printable (2 layers)	/	No
Complex load [MUK 07a, MUK 07b]	NA	5 × 12 mm²+20 × 20 mm²	UWB	Printable	/	No
Patches with stub [BAL 09b]	3 bits	125 × 50 mm²	2.15–2.45 GHz	Printable	/	No
Multi-dipoles [JAL 05a, JAL 05b]	5 bits	25 × 30 mm²	5.45–585 GHz	Printable	2 m (500 mW)	No
Layered patches [MUK 09]	2 bits	18 × 18 mm²	UWB	Printable	/	No
Peano/Hilbert [MCV 09]	5 bits	150 × 30 mm²	0.7–0.9 GHz	Printable	/	No
Slot dipole [BLI 09]	3 bits	NA	UWB	Printable	/	No
Multi SRR [JAN 10]	4 bits	10 × 20 mm²	8–12 GHz	Printable	Contact	No
THz Tags [TED 10, BER 11]	15 bits	NA	THz	Printable (multi layers)	Contact	No
SAR Imaging [INK 16, HAR 09]	96 bits	108 × 15 mm²	60 GHz	Printable	2.5 m	Spatial
TFTC [SUB 05]	4 bits	80 × 50 mm²	13.56 MHz	Printable (multi layers)	< 1 m	Yes

Table 2.2. *Comparison of printable and chipless RFID technologies*

2.3. Comparison of current chipless RFID technologies

We can now compare the main chipless RFID technologies which have been proposed to this day and most of which are still at the research stage. Table 2.2 gathers all technologies mentioned in section 2.1. The performance criteria of the tags defined here are storage capacity, read/write capacity, necessary surface, required bandwidth, production cost, reading distance and collision management. The highlighted rows in this table represent fully potentially printable technologies with simple processes which allow us to achieve the lowest manufacturing costs. Unfortunately, there is no universal chipless solution which allows us today to obtain both a large storage capacity and a reduced surface at a moderate cost. However, some solutions offer a compromise and we have selected a few operating in UWB or lower frequency ranges.

Before listing the different current and future applications of chipless RFID, we should compare their characteristics with those of the optical barcode and conventional RFID. In comparison to the optical barcode, chipless RFID shows:

– a lower coding capacity;

– a similar production cost if conductive ink printing techniques are used;

– an improved reading robustness;

– a superior reading range.

In comparison to conventional RFID, chipless RFID:

– is considerably less expensive, especially if printing techniques based on conductive ink are used;

– is more resistant to harsh environments (nuclear radiation, extended temperature range and electrostatic charges);

– is more resistant to mechanical stress (no bonding/welding of the antenna on the chip);

– can be detected most of the time at a smaller reading distance (approximately, 1 m);

– has a much more reduced storage capacity;

– has frozen and non-rewritable data;

– has no mechanism for collision management.

Therefore, we can see that chipless RFID technology has some of the characteristics of both RFID and barcodes. It thus incorporates some of the benefits of both technologies. In the next section, we will see which are the current and future applications that use or are likely to use chipless tags very soon.

2.4. Market study on printable and chipless RFID technologies

The market share of chipless RFID today is extremely low compared to HF and UHF RFID markets. This is somewhat less true if we consider anti-theft systems such as 1-bit chipless RFID tags. The emergence of studies on chipless RFID is driven by the idea of reducing the unit cost of RFID tags, in order to mainly address the global identification market of consumer goods which today is dominated by optical barcodes. The post and library markets for the identification of parcels and books are potential opportunities. However, a unit cost of 1 dollar cent [HAR 09] is the necessary condition set by manufacturers, in order to be able to address these markets. In general, what can bring RFID technology in an identification field dominated by barcodes is primarily the addition of traceability features, as well as a more robust and remote reading capability. These characteristics open the field to automatic identification systems. In the years to come, a lot of hope is being placed on printable chipless RFID technology, so that it can dethrone the optical barcode or at least, compete with it. The following study is based on a few results from a market study carried out by IdTechEx in 2009 on the growth forecasts of printable and chipless RFID by 2019 [HAR 09].

The number of chipless RFID tags sold in 2009 amounted to more than 2 billion compared to 30 million for printed and chipless technologies. However, in 2009, it was expected that the trend would be reversed in the next few years and that there would be a net benefit of printed and chipless technologies in 2019 with 533 billion units sold compared to 140 billion for chip technologies. Paradoxically, the market shares of printed and chipless RFID tags should represent a third of the overall budget with 4 billion compared to 8 billion for tags with chip. This is, in particular, explained by the decrease in the chipless tag cost which should reach an average of 0.4 euro cents in 2019 compared to 4 cents for tags with a chip. Figures 2.16(a) and (b) show the evolution of units sold and of the market shares for RFID technologies with and without a chip announced in 2009 [IDT 16]. It is clear that currently these forecasts have been very optimistic compared to the market reality. We are still quite far from the announced figures, and even if the potential of this technology is recognized in this study, currently chipless RFID is still at the research and development level, mainly in laboratories. In addition, these quite large figures must be compared to the 10,000 billion barcodes created each year in the world.

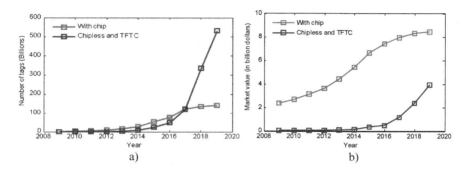

Figure 2.16. *Growth forecasts of RFID technologies up to 2019 – study published in 2009 [IDT 16]: a) in units sold and b) in market value [HAR 09]. For a color version of this figure, see www.iste.co.uk/vena/chipless.zip*

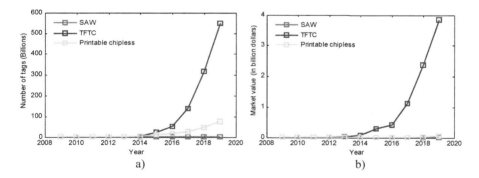

Figure 2.17. *Growth forecasts in different printable and chipless RFID technologies up to 2019: a) in units sold and b) in market value [HAR 09]. For a color version of this figure, see www.iste.co.uk/vena/chipless.zip*

If we are particularly interested in chipless technologies, we can make a comparison. The most important families are non-printable SAW tags, printable TFTC tags and printable chipless tags.

The largest market share predicted for 2019 is attributed to TFTC technologies (printed electronics) with 549 billion units sold, as shown in Figures 2.17(a) and (b). The main reasons are the ability to replace tags with chips without having to change the reader, and to maintain the functionality of read/write access, as well as the ability to disable the tags. Again, currently, we are not anywhere near the technological processes that allow us to obtain such a performance. However, manufacturers are ready to pay more money in comparison to the production cost of a barcode, in order to benefit from the advantage of certain features, such as the features of RFID tags with silicon chips. However, there is some uncertainty about these very prospective and clearly optimistic figures, which, for example, ignore the fact that the technology is still nowhere near mature enough for industrialization.

SAW technology, which was the first marketed chipless tag solution, will remain on very specific markets and, for 2019, the number of sales is estimated at nearly 200 million units.

Printable chipless technologies will address many niche markets (with forecasts of lower amounts than for TFTC tags and 75 billion units planned to be sold in 2019). In this low-cost tag category where we find printed chipless RFID tags, the main studies that will need to be carried out in the following years are:

– the definition of a standard to create open identification systems as for the optical barcode;

– to ensure that the tag and reader costs are lower than those of RFID (in any case in terms of tag unit costs), all compatible with one of the production processes at a very large volume and undoubtedly already existing and widespread;

– the increase in the storage capacity;

– the ability to write data or at least the ability to disable the tags;

– to ensure reading flexibility greater than that of the barcode, i.e. compatible with completely automated information collecting processes and without knowing *a priori* the environment or the object on which the tag is placed.

Under these conditions, printed chipless tags will be able to be fully established compared to the two giants: RFID with chips and barcodes.

2.4.1. *Current applications*

Current applications of chipless RFID technology are still poorly identified, with the exception of certain applications such as the previously mentioned EAS anti-theft systems which address colossal markets. Anti-theft systems in fact use the simplest version of chipless tags, i.e. 1-bit tags operating in HF frequencies [TAG 16] (loop antennas) and microwaves (a dipole connected to a diode). Annual sales amount to 6 billion euros and the unit cost of a tag is approximately a few cents. This mass market is, therefore, an example to follow and all the conditions are met to repeat this success with multi-bit printed chipless tags. This is possible on the condition, of course, that we propose reliable and low-cost reading systems.

Chipless tags based on SAW technology by RFSAW [RFS 16] have also encountered success, but at a much smaller scale compared to EAS systems. In this case, it is not their cost which has allowed their deployment, but rather their ability to operate in harsh environments, as well as the fact that their reading range is greater than the range that can be obtained with passive RFID tags. In fact, SAW tags are able to operate in extended temperature ranges ($+/-200°C$) compared to those with chips (-50 to $+85°C$). In addition, they have a certain resistance to gamma radiation, which makes them compatible for use in the space field. National Aeronautics and Space Administration uses these tags to facilitate the storage of objects in the international space station. In the field of food import/export, sterilizations are performed with the use of gamma radiation, which generally destroy the tag chips. Studies have shown that SAW tags are not damaged by this sterilization phase [HAR 09], which opens the way to the traceability of goods at a global level. SAW tags by MicroDesign are also used for the control of vehicles at automatic motorway tolls. Their large reading range and their lower unit cost compared to the cost of active RFID tags have enabled them to be established on the highways of Norway.

Overall, current and potential markets for different chipless RFID technologies marketed today are very specific niche markets, hindered mainly by the lack of a standard and, for some, by the tag price. These criteria are fundamental to addressing more global markets.

2.4.2. *Future applications*

With the ongoing developments, particularly on the conductive ink printing processes on paper or plastic substrates, which will probably enable us to reach costs of approximately 0.4 cents in 2019 [HAR 09], new prospects are possible. For example, the study presented in [HAR 09] assesses that a capacity of 24 bits will be necessary to address large markets and that 128 bits will be required to address consumer goods markets.

With chipless tags, the expectation of companies specialized in the shipping of goods is not limited to the identification of pallets, but even the identification of each product can be achieved. This feature is currently not implemented with optical barcodes, where a code identifies a product family and not a single product. This feature is eagerly awaited, in fact, it would provide a new dimension in terms of traceability. Each product could thus be tracked independently at each shipping step.

In addition, chipless RFID tags, like tags with chips, can transform into sensors (see Chapter 6). Thus, low-cost tags associated with fragile products such as food may provide, for example, indications on the continuity or not of the cold chain. A generalization of the use of chipless RFID tags will be possible only if the tag cost becomes similar to that of a barcode, with the limit being set at a cost of less than 1 dollar cent. This kind of market will be addressed with chipless tags printed directly on products. Currently, it is dominated by barcodes and several thousands of billions of units are sold per year. However, as mentioned above, a storage capacity of 128 bits is the minimum required to address this market.

Chipless tags also have a role to play in the fields of transport and access control. Today, almost all major cities of the world have an urban transport network which includes automated control systems involving RFID technologies. In this market, HF RFID technology is established, with millions of tickets sold each year for larger networks. The unit cost of an HF RFID tag is approximately 40 euro cents. Therefore, there is a real interest in considering the use of chipless RFID tags as a replacement for these technologies. To address this market, the main obstacles which have to be overcome for chipless RFID are the ability to write data in the tag, as well as the greater storage capacity which should be increased to 128 bits. Again, the costs must remain very low so as to be able to ensure the application of disposable tickets, i.e. with a virtually single use, for the replacement, for example, of magnetic strip tickets.

Finally, postal parcel shipping centers are forced today to use 1D or 2D barcodes for object identification. The main reason for this is the extremely high cost of active RFID tags. However, RFID systems are much better adapted to automatic identification than optical barcodes, in particular due to a certain reading robustness, at large distances, provided by RF waves. Thus, it is clear that chipless RFID technology has a role to play in the field of parcel shipping. This market is huge, as 650 billion objects were mailed in 2009 by the US Post. In France, in 2010, according to a study by ARCEP [RFS 11] (Regulatory Authority for Electronic and Postal Communications), the number of objects mailed amounted to almost 19 billion.

2.5. Issues covered in this book

The two previous sections have allowed us to compare different existing chipless RFID solutions and to determine the required performance according to the target markets. We can, therefore, note that for the most part, a considerable difference still separates the performance achieved from that expected, for a real boom on the industrial plan. In this book, we will present some advances which will help us answer some of the current limitations of chipless RFID technology:

– according to the comparison carried out, it is clear that chipless tags which are based on an information coding in the frequency domain are approaching most of the technical requirements in application terms. That is why the studies described in this book focus on this tag technology. In addition, the potentially printable aspect of several designs of this type which are proposed in the latest developments is a strong argument. For this reason, we exclude from this study SAW tags, TFTC tags and tags based on a temporal coding approach;

– a first study, but not one of lesser importance, will be to search for solutions to reduce the surface of chipless tags. Today, the design proposing the greater storage capacity, i.e. 35 bits, requires a large surface (approximately 88×65 cm^2), i.e. a size greater than that of a credit card (85×55 cm^2);

– we will then present studies on the improvement of information coding, in order to increase the storage capacity of tags. We have seen previously that the minimum requirement is 24 bits;

– a study will be conducted to attempt to improve the detection robustness of chipless tags. Thus, we will address the issues of tag positioning, polarization, reading range and disturbance related to its surrounding environment. This study is particularly important because it has never really been carried out by the scientific community, although the disturbance of the environment on the tag remains one of the current limitations of chipless RFID tags;

– according to the latest developments in the chipless RFID field, the proposed reading systems are almost non-existent and the systems mentioned above are far from meeting the requirements of the regulatory authorities such as the FCC in the United States or the ETSI in Europe. We will attempt to answer this, in particular, with the development of a UWB detection system based on impulse radio transmission.

2.6. Conclusion

This chapter has introduced different chipless RFID technologies, which, as in the conventional RFID field, are numerous and quite diverse. A brief study of current and future markets has enabled us to establish the necessary conditions, in order to ensure that chipless tags could be established as a competitive and reliable identification technology. The necessary conditions are low production cost, increased detection robustness, minimum storage capacity of 24 bits (ideally 128 bits) and the ability to potentially add other features such as the environmental sensor concept. From that point, the already existing chipless technologies have been identified and compared according to several strategic criteria which are essential to their mass deployment. These key points will be addressed in the following pages of this book. Chapter 3 outlines the different coding techniques used in chipless RFID and provides directions to increase coding capacities for a given tag surface.

Information Coding
Techniques in Chipless RFID

3.1. Introduction

Information coding is an essential issue in chipless RFID technology. It directly affects the performance of a chipless tag. This chapter analyzes the latest developments on the different coding methods that can be employed in chipless RFID technology. A general introduction on how to assign an ID to any signal will precede a more detailed study on the techniques employed in the case of a temporal or frequential approach. Performance criteria will be introduced, in order to compare the different possible solutions.

3.2. Waveform and informational content of a signal

For a unit identification of an object, we can simply assign a unique code. For this, the physical principles based on the use of electromagnetic waves both at optical frequencies and radio frequencies can be used. For example, each piece of hardware can be remotely identified to the extent that it generates a signal, with a clean electromagnetic signature when it is illuminated by an incident wave. In this case, we refer to a radar echo, or Radar Cross Section (RCS). In the framework of a conventional RFID tag, the approach is different and the exchange of information is performed in both communication directions with the use of an amplitude-modulated sine wave carrier. The information is linked to the modulation of the signal.

In Figure 3.1, on any signal, a Y dimension as a function of X is represented by different elements that can be used to assign a code to the signal. We see how it is possible to distinguish this signal from other signals of the same type. Indeed, each irregular change in the amplitude such as maximum, minimum, oscillation, or slope variation, can be exploited, i.e. it can be quantified to ultimately be translated, for example, into a binary code. We note that Figure 3.1 describes a signal at one dimension. In reality, the electric field is a vector quantity contained in a plane perpendicular to the propagation direction. Generally, it is a two-dimensional signal that can be used, i.e. the projection of the electric field on two perpendicular axes (horizontal and vertical component). Similarly, this signal can be expressed as a function of time or frequency. In this second case, the signal is complex and can thus be expressed in terms of amplitude and phase. Both quantities can potentially be used for information coding.

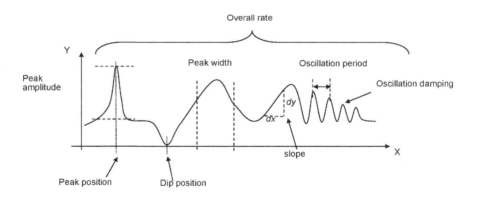

Figure 3.1. *Any signal containing exploitable features to assign a code which allows the identification of this signature*

Simply note that the absence or the presence of a peak or a dip for a given position on X axis can code binary information. So, we refer to ON/OFF (on/off keying (OOK)) or amplitude (ASK) coding. If this same peak is always present and can be positioned on different points on X axis, coding is performed in position. Coded information is no longer binary since several positions can be defined in a given window. In this case, we refer to pulse-position modulation (PPM).

Modulating peak width is also a way that can be used to code information in a signal. We refer to pulse-width modulation (PWM). Finally, a coding based on the modulation of an oscillation period following X (FSK) or of its phase (PSK) can be considered.

3.3. Basic principle of coding

3.3.1. *Presence or absence*

This type of coding is quite widespread, as its implementation is easy and it is robust. It consists of the modulation of the presence or absence of remarkable elements, or symbols, of a signal such as a dip and a peak at the given X positions. The robustness of this coding is related to the fact that peak amplitude will not modify the code. It is sufficient to attach two very distant levels for the assignment of 0 or 1.

For a given X window, we can assign independent k symbols. This provides a number of relatively large N combinations. This number increases to the power of k as shown in equation [3.1]. However, in this case, the coding efficiency is low, with 1 bit per symbol.

$$N = 2^k \hspace{4cm} [3.1]$$

Figure 3.2 shows this coding principle. In this example, a capacity of 3 bits is reached.

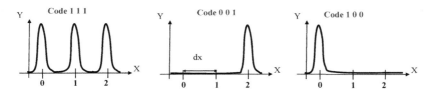

Figure 3.2. *Coding principle in the absence/presence of the symbol (OOK) k=3, n=9 or 3 bits*

3.3.2. *Pulse position coding*

PPM coding consists of the variation of the position of a symbol, usually a peak, in a given X window. Thus, a peak can take several positions. As shown in Figure 3.3, the amount of information that it contains is, therefore, larger. However, it requires a larger spread on X for a given resolution. This code is also robust, because a peak is always present and a variation on its amplitude does not modify the information that it contains.

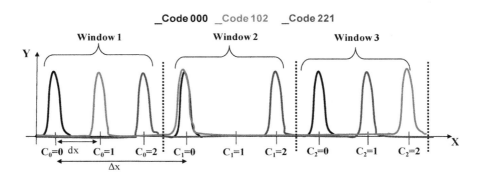

Figure 3.3. *Position coding principle. ΔX = 3, dX = 1, k= 3, N= 9. For a color version of this figure, see www.iste.co.uk/vena/chipless.zip*

Equation [3.2] can be used to estimate the coding capacity of such a device:

$$N = \left[\frac{\Delta X}{dX} \right]^{k} \tag{3.2}$$

In this equation, dX is the resolution (spacing between two peaks) and ΔX is the window in which the peak can be positioned. In addition, if we consider the "absence of peak", in this case [3.3] must be described in the total number of achievable configurations:

$$N' = \left[\frac{\Delta X}{dX} + 1 \right]^{k} \tag{3.3}$$

We note that in this case, it is possible to associate more than one bit per symbol, which makes this coding more effective when the symbol number is limited. In practice, a symbol (peak or dip) will most of the time correspond to a resonator, and for a given surface, the resonator number is inevitably limited. In this way, this technique will allow us to increase the coding density per surface.

3.3.3. *Coding on symbol width*

Modulating symbol width is also a possible means that can be used to code information in a robust way. Sometimes, we refer to PWM.

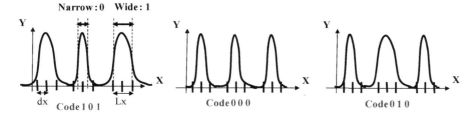

Figure 3.4. *Symbol width coding principle,*
Lx = 2, dx = 1, k= 3, N= 8.

In the graphs presented in Figure 3.4, an example of coding of this type, for 3 symbols, is presented. Each symbol may vary depending on two possible widths, thus coding binary information. If the width can take more than two values, then in the same way as for a coding in a symbol position, the number of possible combinations can be calculated using [3.4], with Lx, the window in which the peak can be spread out, and dx, the resolution:

$$N = \left[\frac{Lx}{dx} \right]^k$$

[3.4]

3.3.4. *Coding on signal waveform*

A signal can also be modeled by the whole of its evolution versus X and not only by the values at specific positions of X, such as a peak and a dip.

We can, therefore, assign a code to a given signal waveform (see Figure 3.5). In this case, the number of combinations is limited by the resolution in X and Y, and by the windows observable in X and Y.

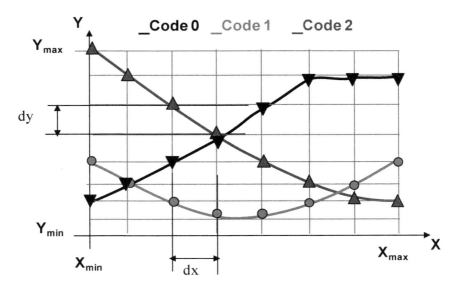

Figure 3.5. *Coding principle according to the recognition of the signal waveform. For a color version of this figure, see www.iste.co.uk/vena/chipless.zip*

Thus, the formula [3.5] allows us to obtain the amount of information that can be achieved with this type of coding

$$N = \frac{(Y_{max} - Y_{min})(X_{max} - Y_{min})}{dy \cdot dx}$$ [3.5]

In practice, all combinations described by equation [3.5] are difficult to obtain, because certain signals which are to be generated are hardly feasible and may not correspond to any physical signal reflected by a tag. For example, if we use multiple resonators in parallel to generate a particular signature, we will be limited by the quality factor of each resonator, which reduces the minimum width of the peaks. All values, $Y_{max}, Y_{min}, X_{max}, Y_{min}, dX, dY$ are, therefore, also directly related to the performance that it will be possible to obtain from the resonators chosen for the tag design.

3.4. Temporal coding

After generally describing the major coding principles, we are now going to discuss the constraints related to the physical elements that will be used for the tag creation. Thus, in this section, we will describe one of the two major families of chipless tags, namely, those that use temporal coding. Compared to what has been introduced previously, the noted X axis here represents the time. From that point, we obtain in a very understandable manner a definition of a temporal tag, i.e. a tag where its information will be retrieved from the temporal expression of the backscattered signal versus the time. All coding techniques described above can, therefore, apply if it is physically possible to create a tag with the desired properties. In practice, due to the inherent limitations in chipless technology in terms of cost, dimensions, frequencies, etc., only certain coding approaches are used, as we will see below.

Another important point is that it is essential to understand that the coding described in this chapter, and more accurately in this section, is independent of the reading mode used to query the tag and retrieve its ID. Thus, a temporal tag (or a frequency tag, as we will see in the next section) can also be read with a reader transmitting a short pulse (temporal reading approach) than a continuous wave signal of which we will vary the frequency. Regardless of the chosen reading method, it is at the postprocessing level that the distinction takes place. For example, in the second case, it will be necessary to use an inverse Fourier transform to trace the temporal signal and it is on this signal that the tag ID will be decoded.

In conventional communication systems, the evolution of the signal as a function of time is used for information coding. This is particularly the case for the conventional RFID which uses amplitude and phase modulations as a function of time.

Regarding chipless RFID technology, in the temporal domain, the design of identification devices is based primarily on the concept of reflections between different environments. This is reflected in the field of microwaves by the presence of discontinuities or of characteristic impedance variations throughout the wave propagation, at the tag level. Thus, a rudimentary approach is to place a number of discontinuities at different distances, in order to obtain a signal whose information is coded by the temporal positioning of reflections. These discontinuities can be lumped [ZHA 06] or distributed [ZHE 08] capacities. The coding used can be performed either in the absence/presence of pulse (OOK), or in PPM. Contrary to a conventional transmission system, in the case of chipless RFID tags, the configuration is set, therefore, the temporal signature extracted from the tag is always the same regardless of the query time. This coding principle is extremely simple to implement and it is also at the origin of the first chipless tags that use SAW technology. In the case of temporal coding, the coding capacity is related to the numbers of reflections, to the temporal resolution and to the temporal observation window (see section 3.3.2). The main limitation of this type of coding is due to losses arising on the incident wave guided in the structure, particularly due to reflections. Figure 3.6 illustrates the temporal coding principle in the OOK approach.

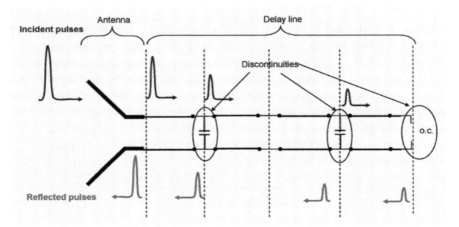

Figure 3.6. *Temporal coding principle – OOK. The level of reflected pulses is attenuated after each new reflection. For a color version of this figure, see www.iste.co.uk/vena/chipless.zip*

3.4.1. *Presence/absence or OOK*

OOK coding consists of modulating the presence of a symbol over time. The structure presented in [ZHA 06] uses this coding principle. It allows us to obtain a capacity of 4 bits, for a size of 8.2 × 3 cm². It is composed of a coplanar transmission line, on which discontinuities, corresponding to shunt capacitances, are placed. Figures 2.3(a) and (b) in Chapter 2 show the structure studied. Depending on the presence of these discontinuities, reflected waves are created and retransmitted to a receiver. The presence of a reflection can be regarded as a logic state of "1", while the absence of reflection will be viewed as "0". The duration of a symbol must be detectable by the reading system, which will set the temporal resolution $dX=dt$. From [3.1], we can obtain the number of combinations [3.6] by replacing k with $\Delta t/dt+1$ and $\Delta X = \Delta t$, the observation window.

$$N = 2^{\frac{\Delta t}{dt}+1}$$

[3.6]

Let us take a sampler at 40 Gs/s. Consider a temporal resolution of 25 ps. To properly recognize a pulse, we will set a resolution of approximately 100 ps. For an observation window of 1 ns, we can, therefore, see 10 pulses, i.e. $k = 10$ or 10 bits. And for 6.4 ns, we can code 64 bits. Here are the dimensions of the lines corresponding to a duration of signal propagation of 6.4 ns (to achieve 64 reflections) as a function of the permittivity of the materials used:

– air ($\varepsilon r = 1$) : 192 cm;

– FR-4 ($\varepsilon r = 4.6$) : 89.5 cm;

– ceramic ($\varepsilon_r = 50$) : 27.15 cm.

We observe with this example that the line lengths which are necessary to code the information in this way become large very. If we add to this the attenuation related to the presence of each reflection, we come to the result that a very small number of bits (approximately a half-dozen) can be coded in this way on chipless tags consisting of a conventional RF line.

3.4.2. *Pulse-position or PPM*

PPM coding consists of changing the position of a pulse in a given temporal window. This type of coding requires a more extended temporal span than an OOK coding for an equivalent capacity. By contrast, it has the advantage of requiring a lower number of reflectors. This is reflected by lower insertion losses. In fact, each discontinuity is at the origin of a reflection and, therefore, induces losses. SAW tags by RFSAW use this coding principle [HAR 02], in the same way as the tag by Vemagiri *et al.* [VEM 07], for which a graphical representation is provided in Chapter 2, Figure 2.4(a).

Figure 3.7 represents the coding capacity in bits as a function of the number of reflectors for an observation window of 10 ns and a temporal resolution of 100 ps. In the case of a conventional PPM coding, from [3.2], we can establish the formula [3.7] with $\Delta X = \Delta t$ and $dX = dt$. For the extended PPM coding in the case of a peak absence, in the same way from [3.3], we can obtain the formula [3.8].

$$N = \left[\frac{\Delta t}{dt} \right]^{k} \tag{3.7}$$

$$N = \left[\frac{\Delta t}{dt} + 1 \right]^{k} \tag{3.8}$$

We can observe two limiting cases:

– 1 pulse in the temporal window of 10 ns: in this case, we can code 100 positions or 6.6 bits;

– 100 pulses in the temporal window of 10 ns: we have an OOK coding with 100 bits for 100 pulses;

In Figure 3.7, we can clearly see that the extended PPM coding increases the coding capacity of 50–64 bits for 32 reflections.

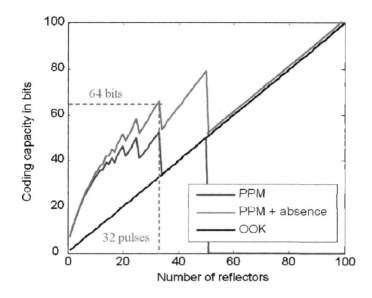

Figure 3.7. *Coding capacity as a function of the number of reflectors for a resolution of 100 ps and an observation window of 10 ns. For a color version of this figure, see www.iste.co.uk/vena/chipless.zip*

3.4.2.1. *Phase jump or QSPK*

To increase the coding efficiency of the previously presented delay lines, a possible solution used by Mandel *et al.* [MAN 09] is to modulate the phase of the reflected pulses depending on several states. In this study, which has been partially presented in Chapter 2, a CRLH line allows us to reduce the line length which is necessary to separate the pulses in time. Instead of using the parasitic capacitances to create reflections, complex loads of different values are used in order to change the phase of the reflected pulses, as shown in Figure 3.8. Thus, with four different loads, it is possible to obtain four distinct-phase values. This technique is similar to a quadrature phase shift keying (QPSK) coding, which is widely used in conventional communication systems. The total coding capacity is provided in equation [3.9]. In the example shown in Figure 3.8, with four reflectors, we obtain the code "2.1.3.0". The total ID number that can be obtained with $k = 4$ reflectors is $N = 256$ or 8 bits.

$$N = 4^k \hspace{4cm} [3.9]$$

With $k = 5$ reflectors, the coding capacity will be 10 bits. In the case of a temporal coding, it is very interesting to implement such an approach, because in this case, the factor which will limit the total information capacity N is the maximum number of reflectors k. In fact, to increase the number of reflectors implies inevitably an increase in the line length and an increase in insertion losses.

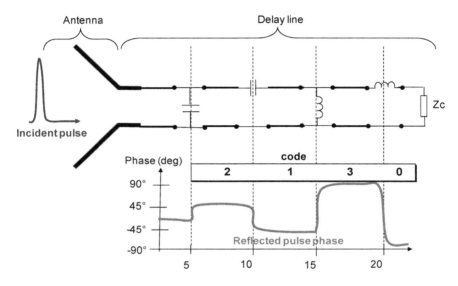

Figure 3.8. *Temporal coding principle – QPSK. The code generated by the reflectors is "2.1.3.0". For a color version of this figure, see www.iste.co.uk/vena/chipless.zip*

3.4.3. *Anti-collision principle*

Coding in the temporal domain allows us to implement anti-collision principles. Contrary to the conventional RFID, whose collision management is based at the communication protocol level, in the case of a chipless RFID, we have to ensure that the responses of different tags are separated temporally and/or frequentially.

SAW tags use several anti-collision principles. The first is based on a temporal separation [HAR 04]. In fact, these tags are designed in such a way that they can respond in several temporal windows or "slots" (see Figure 3.9(a)). This parameter is set during the tag manufacturing process.

They use piezoelectric substrates which transform an electromagnetic wave into an acoustic wave that is propagating very slowly. This allows us to obtain considerably large delays. A large number of slots are, therefore, possible on a compact surface. We can also add that if the tags are sufficiently separated in space, and that their distance with respect to the antennas is different, their electromagnetic responses can be separated without using a supplementary delay line. This principle allows us to read simultaneously tags placed very close to each other.

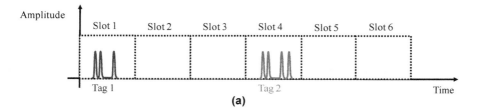

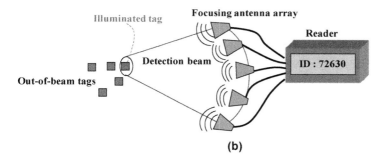

Figure 3.9. *a) Anti-collision principle based on the addition of a delay before the response reflection; b) spatial anti-collision principle using a focused antenna array. The tags are sufficiently separated in space, so the reader detects only a single tag at the same time. For a color version of this figure, see www.iste.co.uk/vena/ chipless.zip*

Another anti-collision technique involves the use of narrow-beam antennas. In this way, in a group of adjacent tags, only one will receive the incident wave and retransmit its signature. To reduce the detection beam size, a focused antenna array (see Figure 3.9(b)) can be used. Thus, a focal point of approximately 12 cm for a distance of 1–2.2 m is possible [HAR 04].

Another important coding aspect regards the ability to implement error corrections. Thus, coding can be chosen in such a way that an error on one or several bits which is related to a collision can be corrected. For that, we use error-corrector codes. Thus, linear codes are used in [HAR 04] to improve the anti-collision system of the readers of SAW tags.

Another anti-collision principle which can be used to read multiple tags simultaneously is to ensure a sufficiently large spatial separation [PER 14]. This design is used in the solution provided by Inksure [INK 16], operating on frequencies of approximately 60 GHz.

3.5. Frequency coding

3.5.1. *Amplitude rate*

In the frequency domain, it is possible to code information by focusing on amplitude frequency variations of the wave retransmitted to the reader. Such studies have been carried out by placing resonant items near a line [PRE 09b, PRE 09a, PRE 08] or by altering the resonance frequency of dipoles in networks [JAL 05a, JAL 05b]. Possible codings are similar to those used in the temporal domain with the difference that we will no longer refer to pulse in time but to resonance in spectrum. Parameter X, which was defined at the beginning of the chapter, here is frequency. An OOK coding is reflected in this case by the absence/presence of resonance for a given frequency, as shown in Figure 3.10(a). In this example, half-wave resonators such as those of dipoles in short-circuit are used. They are excited by a continuous wave where a frequency scanning between 4 and 7 GHz is performed. For a PPM coding, the resonance frequency shift will allow us to obtain the information.

3.5.1.1. *Presence/absence of OOK*

In most cases, the coding used is of the following type: presence/absence of resonance in the spectrum. This is an OOK coding transposed in the frequency domain. Thus, a resonance is equivalent to 1 bit. In the example in Figure 3.10(a), three dipole resonators in short-circuit are visible, generating three resonance peaks. The resonator associated with the 6 GHz frequency is not present. This gives the code "1.1.0.1". From [3.1], we can deduce formula [3.10] to calculate the coding capacity of such a device:

$$N = 2^{\frac{\Delta f}{df}+1}$$ [3.10]

This formula is very close to [3.6], with the difference that we are referring to frequency resolution df and available frequency window Δf.

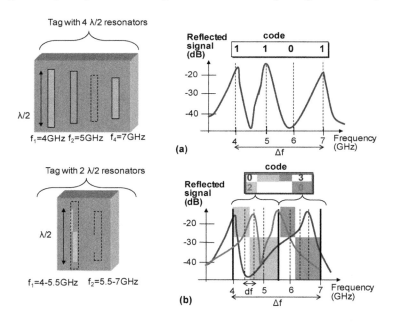

Figure 3.10. *a) OOK frequency coding principle with a tag with four dipole resonators in short-circuit, Δf=3 GHz, df=1 GHz, N=16. The generated code is "1.1.0.1"; b) PPM frequency coding principle with a tag with two resonators, Δf=3 GHz, df=0.375 GHz, N = 16. For a color version of this figure, see www.iste.co.uk/vena/chipless.zip*

In [PRE 09b], a tag with 35 coded resonators in the absence/presence of resonance is presented. The frequency band used is between 3 and 7 GHz, i.e. a band compatible with UWB. Spectrum measurement in open space allows us to reach minimum resolutions of approximately 50 MHz. Thus, the coding capacities that can be achieved as a function of the allocated bandwidth can be:

– 2.5–7.5 GHz: 101 bits;

– 3–9 GHz: 121 bits;

– 4–12 GHz: 161 bits.

In practice, these coding capacities require the presence of a large number of resonators, which increases significantly the tag surface (a bit for a resonator). In addition, the insertion losses of the multi-band notch-filter used in [PRE 09b] limit the number of resonances that it is possible to create in the spectrum. Finally, certain resonators generate higher order modes that can interfere with the resonances located at higher frequencies. It is more difficult in practice to ensure a resolution of 50 MHz on all of the above mentioned bands. In fact, between 5 and 6 GHz, a resolution of 100 MHz should be considered.

3.5.1.2. *Frequency hopping or PPM*

In the frequency domain, a PPM coding has the great advantage of reducing the tag size compared to an OOK coding for an equivalent coding capacity. In fact, in the case of a conventional OOK coding, a resonance is equivalent to a resonator and, therefore, only to 1 bit. The capacity in bits increases linearly with the number of resonators. In the case of a frequency shift coding, we can associate a resonator with several bits, as shown in Figure 3.10(b). However, for an equal coding capacity and an equivalent frequency resolution, we observe a spread in the spectrum. From this point, it is interesting to introduce the concepts of density of coding per surface unit (DCS) and spectral density coding (SDC). In the case of PPM coding, DCS is improved at the expense of SDC. Equations [3.11] and [3.12] can be used to calculate coding capacity, respectively, in the case of the use of a PPM coding and of a PPM coding with the absence/presence of a symbol. It should be noted that simply by adding a supplementary state in [3.12], we can considerably increase the coding capacity for a number of identical resonators.

$$N = \left[\frac{\Delta f}{df} \right]^{k} \tag{3.11}$$

$$N = \left[\frac{\Delta f}{df} + 1 \right]^{k} \tag{3.12}$$

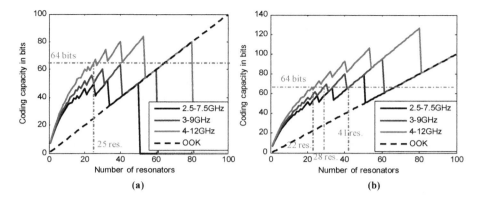

Figure 3.11. *Coding capacity for a resolution of 50 MHz:*
a) PPM coding and b) PPM coding + presence/absence.
For a color version of this figure, see www.iste.co.uk/vena/chipless.zip

Let us use the above bandwidths to estimate the achievable coding capacities as a function of the number of resonators. We consider that frequency resolution is always set at 50 MHz. Figure 3.11 presents the achievable coding capacities in the case of a conventional PPM coding (Figure 3.11(a)) and in the case of an extended PPM coding with absence of resonance (Figure 3.11(b)). We note that a coding capacity of 64 bits is achievable with 28 resonators in the case of an extended PPM, for a frequency band of 3–9 GHz. PPM coding has been used for the design of miniaturized chipless tags [VEN 11a] with a capacity of 10 bits using three resonances for frequencies ranging from 2 to 5.5 GHz. This design is described in Chapter 4.

Finally, it is possible to further extend the coding capacity with the same constraints (number of resonators, frequency range, etc.) by not further restricting the resonance position of a resonator in a given sub-band Δf. In fact, if we consider that Δf is equal to the whole available band and that frequencies of k resonators can be independently distributed by a step of df on the whole band, coding capacity is increased in a consistent manner as shown in [HAM 13] and [PER 13]. However, in this case, the consideration is that several resonances can be found at a close distance (spaced by df) from each other, which imposes supplementary constraints at the level of the tag and reader design. In fact, it is necessary that the resonators are very strongly decoupled, with large quality factors, to distinguish them from each other and to remove decoding errors. We note that in the previously introduced configuration, there are not many configurations where two resonators can be spaced by df (at worst, two resonators can be adjacent), which limits the negative effects. To the extent that this does not concern many configurations, it is even possible to remove these sensitive cases in such a way as to completely eliminate this difficulty.

3.5.2. Phase

So far, the present coding methods used primarily in chipless RFID are based on modulation techniques of the presence/absence of a peak in the spectrum [JAL 05a]. Nevertheless, it seems that phase coding can provide a better reliability for the detection of chipless tags [PRE 09a, NAI 13]. In addition, a coding density of a larger surface unit can be obtained [VEN 11c]. We will see that a phase coding can even be combined in an amplitude coding to improve its effectiveness and thus reduce the necessary tag surface for an equivalent storage capacity.

3.5.2.1. Phase rate

In a study carried out by Mukherjee [MUK 07a], it has been shown that it is feasible and interesting to code information on the phase variation of the wave as a function of frequency. For this, a phase coding approach based on the use of a complex load placed at the antenna level has been developed. The reading system allows us to trace this complex load as a function of the frequency evolution of the phase. Therefore, the complex load acts as a unique ID. Coding in this case is performed on the overall phase evolution as a function of frequency. In this study, it is shown that four different loads produce four different phase profiles.

To consider the number of possible combinations, we should define a minimum resolution and limit the different parameters of the complex load. These constraints are determined by the manufacturing details of the tag and the reading system. Let us take, for example, the parameters used in [MUK 07a] and the typical values mentioned in different publications on the chipless tags detection systems:

– bandwidth: 2 GHz (5.9–7.9 GHz);

– frequency resolution: 50 MHz;

– phase resolution: 1° (e.g.: component AD8302);

– phase terminals: +/–40°.

From [3.5], we can obtain a total number of combinations n = 3,200 or 11.6 bits. This value is a theoretical limit that can be obtained only if the complex load is able to generate all possible phase rates in the resolution limits and the specified margins. In addition, the use of complex loads is hardly compatible with the technological requirements of the chipless concept.

3.5.2.2. Phase jump

Another design using the phase to code information has been proposed in [BAL 09]. In this case, it refers to modulating the phase shift value for precise frequencies. An analogy can be made with an OOK or PPM coding. The proposed structure was introduced in Chapter 2, Figure 2.10. Three "patch" antennas loaded by a variable-length stub are used. In this example, coding capacity is 1 bit per resonator, because the phase can evolve only between two values (0–35°). However, the structures used are potentially able to code more than two states per resonator if intermediate-length stubs are used. To calculate the number of combinations with this coding principle, we can use equation [3.13] with φ_{max} and φ_{min}, the high and low limits of the phase, $d\varphi$, the resolution of the detection system and k, the number of frequencies used.

$$N = \left[\frac{\phi_{max} - \phi_{min}}{d\phi} \right]^k \qquad [3.13]$$

To conclude, in [VEN 11b], it has been shown that it was possible to use the width of a phase jump in frequency to code data. In fact, "C" resonator presented in Figure 3.12(a) behaves as a wide band phase shifter with a bandwidth that can be modulated as a function of the gap g (see Figure 3.12(a)). The amplitude and phase rate of the reflected signal is presented in Figures 3.12(b) and (c) for two different gap values g. In the simplest case, a narrow bandwidth provides a logic state of "0", while a wider bandwidth provides a logic state of "1". In this study, it has been shown that four widths of bandwidths can be discriminated without error, to trace the tag ID, which leads to a coding efficiency of 2 bits per resonator. This study will be analyzed in Chapter 4.

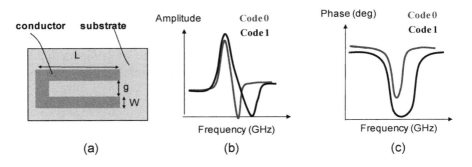

Figure 3.12. *a) "C" resonator; b) frequency amplitude variation and c) frequency phase variation*

3.5.2.3. Group velocity

In many applications, the least possible dispersive structures are recommended, in particular to avoid deforming the transmitted signals. However, it is possible to take advantage of this feature to code information. Thus, it is possible to modulate for each frequency the group delay, which in addition has the advantage of being a more robust physical parameter than amplitude, especially in environments favoring multiple paths and ground clutter [MUK 09, NAI 13]. The group delay sometimes provides information on the duration of a return route of each frequency component of a signal. In this case, we can measure the group delay directly from the temporal response of the tag. Otherwise, it is always possible to calculate it by deriving the phase as a function of frequency as expressed in equation [3.14].

$$\tau_d = -\frac{\partial \varphi}{\partial \omega} \qquad [3.14]$$

In Figure 3.13(a), we can see a transmission line composed of "C" cells (C-sections), which allows us to generate the desired dispersive effect. Figure 3.13(b) presents the coding design with the use of group delay of the output signal of the dispersive line.

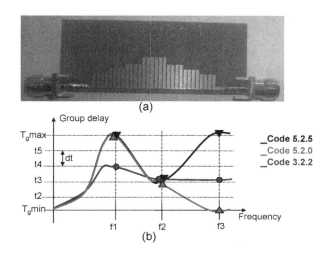

(a)

(b)

Figure 3.13. a) Dispersive transmission line and b) group time coding principle. For a color version of this figure, see www.iste.co.uk/vena/chipless.zip

In this example, three frequencies *f1* to *f3* are used, for each frequency six values of group delay are possible. Thus, the number of possible combinations is $6^3 = 216$ or 7.7 bits. Equation [3.15] can be used to estimate coding capacity.

$$N = \left[\frac{T_g \max - T_g \min}{dt} \right]^k$$
[3.15]

$T_g$max and $T_g$min are the minimum and maximum values of measurable group delays for a given frequency and dt is the temporal resolution. In practice, temporal resolutions of approximately 0.8 ns are possible and group delays ranging between 1 and 10 ns are achievable with C-sections [NAI 13b, NAI 13c]. Resonances, which are related to the number of C-section groups used (see Figure 2.5 where two groups are used), must be separated from almost 1 GHz to guarantee a good mode decoupling [NAI 13c]. Taking these parameters into account, these are the theoretical achievable coding capacities as a function of bandwidth:

– 2–6 GHz: four resonances with five possible values for each ➔ 9.2 bits;

– 3–9 GHz: six resonances with five possible values for each ➔ 13.9 bits.

The studies carried out on this type of structure show that it is difficult to design tags with more than two C-section groups. However, eight different group delay values are easily achievable [NAI 13b]. Finally, coding capacity remains low, i.e. less than 10 bits. However, structures based on this design but created from a coupling surface, and not a side (as those presented in Figure 3.13), allow us to significantly improve performance. Thus, it is possible, from structures created by folding, to keep a very low unit cost, to obtain more than 10 bits on only 250 MHz, i.e. the accumulations of ISM bands at 2.45 and 5.8 GHz.

3.5.3. Anti-collision principle

Studies on chipless RFID tags without chip coding frequency information almost never mention an anti-collision principle [PER 14]. However, in the same way as with temporal coding, the anti-collision principle can be implemented. The simplest solution is to assign to each tag a small part of the usable frequency band. This procedure is relatively robust. However, it dramatically decreases coding capacity. In a way, it is a transposition of the temporal separation solution (see Figure 3.9(b)) mentioned by Hartmann et al. [HAR 04] in the frequency domain.

Other more complex solutions based on this principle exist, particularly for tags where information is coded with the use of group delay [PER 14]. In this case, a double time/frequency discrimination is possible. For this, we also note that the spatial anti-collision principle introduced by Hartmann et al. [HAR 04], which is based on the use of a focused antenna array (see Figure 3.9(b)), can also be used with frequency tags. However, a conventional temporal anti-collision method may be considered. In this case and provided that the tags are sufficiently separated spatially, a time gating will isolate the electromagnetic responses of different tags.

3.6. Coding efficiency improvement

3.6.1. Constellation diagram and graphic representations

To go further in the coding efficiency improvement, we can introduce hybrid coding techniques. "Hybrid" means that coding will be generated by more than one parameter, for example the amplitude can be combined with the phase. In the same way as for the telecommunication systems using conventional modulation principles, a constellation diagram can be established in order to define the coding efficiency for a given frequency, or a given resonator instead of a given time. In the example in Figure 3.14(a), the two parameters used are the amplitude and the phase for a given frequency. This approach is very similar to an IQ modulation scheme transposed to the frequency domain.

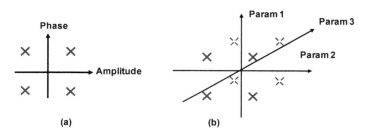

Figure 3.14. *a) Frequency 2D constellation diagram and b) generalization, constellation diagram in three, or even N dimensions*

In certain cases, more than two parameters can be modified for a given symbol (in the temporal or frequency domain), so that a constellation diagram with N dimensions can be adopted to represent N independent parameters, for a given frequency, as indicated in Figure 3.14(b).

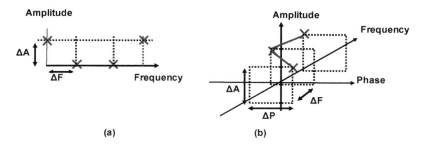

Figure 3.15. *Graphical representation of an ID: a) in 2D, with four resonators, of which only "A" amplitude varies; b) in 3D, with three resonators, of which both amplitude and phase vary*

Previous constellations provide information on the coding efficiency for a given symbol in the temporal domain or more precisely here, in the frequency domain. A variant can be used by integrating the frequency axis (or the time axis depending on the type of coding) to graphically display the generated ID. In this case, the constellation does not provide the total number of possible combinations for each frequency (or time) interval, but the whole of the code spread out over several frequencies (or in time). Such a representation may be useful to compare the different IDs generated by the tags, in order to determine the possible interference. We can imagine that a code recognition technique can be implemented with the use of a graphical method by comparing the constellation modified with the expected responses. However, in the case where increased detection robustness is necessary, a subset of codes can be selected to limit the possible recognition errors between two similar codes. For this purpose, a minimum Hamming distance must be imposed.

For example, the most used conventional coding technique in the frequency domain is to modulate the presence or the absence of a resonance peak in the spectrum. This is similar to an OOK coding and the modified constellation which represents the given code in Figure 3.15(a). Now, if the phase is also used, a modified three-dimensional (3D) constellation must be used to represent the code in Figure 3.15(b). The obtained 3D curve thus represents a unique ID.

Before defining a constellation diagram, we must take into account several practical parameters related to the detection system. The factors limiting the increase in the number of coding states are primarily related to the reading system resolution, and to the noise level generated by the electronics of the reception stage. In concrete terms, a constellation with very adjacent points requires a reading system with a very low noise floor and a very good resolution of the analogue to digital conversion stage. However, a constellation with distant points is more robust in detection, but the coding efficiency of the tag is lower.

3.6.2. *Use of several states*

Most of the coding techniques mentioned in chipless RFID use only one axis of a given constellation and two possible states, as shown in Figure 3.16(a) [FRI 13]. It is, therefore, the smallest possible constellation and consequently, the lowest coding efficiency with 1 bit per symbol. This is the case for many chipless designs which use the "presence/absence" coding technique [3.8] and [3.11]. Nevertheless, this coding is robust due to the strong contrast between the two possible states of each symbol.

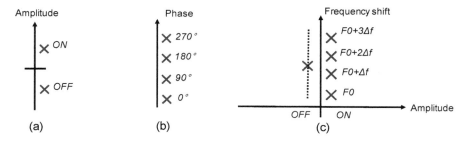

Figure 3.16. *a) Constellation in the frequency domain for the "absence / presence" coding technique; b) constellation in the frequency domain for a QPSK coding; c) constellation in the frequency domain for the coding technique in frequency hopping (PPM) + presence / absence*

The above mentioned studies by Mandel *et al.* [MAN 09] are oriented toward increasing the number of states according to a coding axis. In fact, for each reflector, the phase of the reflected pulse can vary depending on four states, as in the case of a conventional QPSK modulation scheme. The parameter used to code data is the phase. In addition, values between 0 and 270° are possible. Coding efficiency in this case is 2 bits per symbol (see Figure 3.16(b)).

3.6.3. *Hybrid coding*

The above-mentioned coding techniques use a single constellation axis. This means that only the phase, or the amplitude, is used in the frequency (or temporal) domain. However, as we will see here, techniques for improvement of the coding densities of tags particularly via hybrid coding techniques are possible. For example, in the frequency domain, we can control the resonance peak position in a given window, as well as modulate its presence (see Figure 3.16(c)). This coding technique was introduced in 2010 [DEE 10]. It allows us to significantly increase the coding associated with each resonance. As shown in Figure 3.11(b), by simply adding the presence / absence case to different positions that can take a resonance peak in a given frequency window allows us to reach a capacity target of 64 bits with 28 resonators (between 3 and 9 GHz).

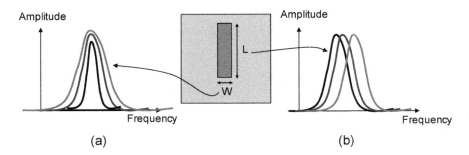

Figure 3.17. *Hybrid coding technique based on: a) the selectivity of a resonator by varying its width W and b) the peak frequency by varying its length L. For a color version of this figure, see www.iste.co.uk/vena/chipless.zip*

Other parameters such as the phase and the damping factor of a resonance peak can be used to enrich the associated constellation diagram. We can imagine, for example, using a dipole in short-circuit (without a ground plane) as shown in Figure 3.17, in order to combine a frequency PPM coding with a resonance peak width coding. For instance, in [VEN 11c] it has been introduced a chipless tag design with multi-resonators that allows us to combine a frequency PPM coding with a coding using phase variations, as shown in Figure 3.12(c). We will analyze this structure in Chapter 4.

Currently, there is no chipless tag design which uses three coding axes for a given frequency. However, intrinsically, as demonstrated by Blischak *et al.* [BLI 09], the analytical response of a radar target can be modeled by a sum of p damped sine waves for p resonant modes [3.16] (or resonant frequency). Each mode has four independent parameters: amplitude A_k, phase Φ_k, damping m_k and pulsation ω_k.

$$h(t) = \sum_{k=1}^{p} A_k \, e^{j\varphi_k} \cdot e^{-m_k t} \cdot e^{j\omega_k t} \qquad [3.16]$$

A design able to vary each of these parameters independently will provide the best coding efficiency that it is possible to have with a chipless tag which can be considered here as a passive radar target. It remains to the reader to be able to recover all this information in an actual environment of use, namely, in conditions where the tag signature can be very low compared to all parasitic reflections, produced by the different objects which are present in the reading scene.

3.7. Comparison of amplitude and phase coding

The detection robustness is an essential aspect of chipless RFID systems, as it allows us to define its performance in particular in terms of reading distance for a given environment. A transmission channel can be disturbed by different phenomena which are listed below.

Medium attenuation: this phenomenon is related to the nature of the objects and mediums traversed by the electromagnetic waves. Its effect, most often, results in the deformation of signals in amplitude and phase. A dispersive medium can even distort the signals in a different way depending on the frequency, which is very complex to predict. The attenuation and dispersion effect does not change over time, unless we consider an environment containing moving objects.

Multiple paths: the wireless communication can be performed according to a direct path (in line of sight) or according to paths generated by the reflections of the waves on the objects. This phenomenon is always present, particularly when the wireless communication systems operate inside buildings.

White noise: this is omnipresent and is due to the thermal agitation of the charge carriers in any electrical conductor. Its effect results in adding a random component to the signals transmitted and received at the detection system level. The amplitude and the phase are strongly affected by the white noise.

Interference with other communication systems: wireless communications are omnipresent. Most of the applications use ISM bands. If a chipless tag operates in ISM bands, the interference with the adjacent channels must be taken into consideration. Most of the communications are quite narrowband compared to the chipless tags which are rather dedicated to operate in UWB. Nevertheless, communication between a tag and its reader can be disrupted essentially around 2.45 and 5.8 GHz.

The effects of the white noise and multiple paths affect the amplitude and the phase. However, certain techniques allow us to eliminate these undesirable effects. For example, to remove multiple paths, we simply have to perform a time gating by considering that the first radar echo arriving at the reception stage is the one that we must take into account. This avoids any destructive interference between the direct path and other paths. Regarding the white noise, increasing the acquisition time by averaging several records enables a strong attenuation.

In addition, the filtering effect of the transmission channel must be taken into account for the communications in free space. It affects both amplitude and phase, but in a deterministic way. In fact, a transmission channel in free space without noise introduces an attenuation effect and a delay on the signals. Expression [3.17] shows the way in which an electromagnetic wave of amplitude $|E(\omega)|$ and phase $\varphi(\omega)$ propagating in free space is affected.

$$\underline{E} = \underbrace{|E(\omega)| \cdot e^{j\phi(\omega)} \times e^{(-\alpha + j\beta)z}}_{channel\ effect} = \underbrace{|E(\omega)| \cdot e^{-\alpha \cdot z}}_{Amplitude} \times e^{\overbrace{j(\phi(\omega) + \beta \cdot z)}^{Phase}} \qquad [3.17]$$

$$\tau_d = -\frac{d(\phi(\omega) + \beta \cdot z)}{d\omega} = -\frac{d(\phi(\omega) + \frac{\omega}{c} \cdot z)}{d\omega} = \underbrace{-\frac{d\phi(\omega)}{d\omega}}_{group\ delay} - \frac{z}{c} \qquad [3.18]$$

We can observe that the amplitude is attenuated as a function of the distance z by $e^{-\alpha z}$ and that the phase ($= j\beta z$) varies linearly as a function of the distance z. Therefore, if we derive the phase as a function of frequency (see relationship [3.17]), we obtain the group delay τ_d of the device, as well as a constant related to the distance z and the wave propagation speed in the medium (C in the air). Therefore, group delay as a function of frequency is not affected, contrary to the amplitude. If we consider the measurement as being instantaneous, z/c is a constant which will not damage the information decoding. We can see that the frequency variation of the group delay is not modified by the distance and, therefore, that the onboard information maintains its integrity until the signal-to-noise ratio become too low. This confirms the experimental results by Preradovic *et al.* [PRE 09a], who have demonstrated that the use of the phase for the tag identification is proving to be more robust, as it allows us to obtain a larger reading distance. They also confirm many of the results [VEN 12a, VEN 12b] which will be presented in the following chapters.

3.8. Coding performance criteria

As we have seen, there are several parameters that we should take into account, in order to define the coding efficiency of a chipless device. The challenge is to code the maximum information in the smallest possible surface and to require the lowest frequency bandwidth. If we consider the data storage capacity, the frequency band and the tag dimensions, it is clear that these three parameters are closely related, even if they describe different aspects. Trade-off must be made between these different performance criteria that change from one application to another. Therefore, it is of interest to establish figures of merit or performance criteria to evaluate the different designs and choose the most suitable as a function of various features that we should take into consideration. That is why as we have seen, it is interesting to introduce the two following criteria:

– the density of coding per surface unit (DCS) in bits per cm²;

– the spectral density coding (SDC) in bits per GHz.

Once these two criteria are defined, we can compare different chipless tag designs which are coded according to a frequency approach. For this, we can establish a representative chart, with DCS on one axis and SDC on the other axis, as presented in Figure 3.18.

① Spiral resonators (8.8cm × 6.5cm), 35 bits, BW = 4000 MHz [PRE 09c].

② Dipoles, (1.8cm × 3.5cm), 5 bits, BW = 400 MHz [JAL 05b].

③ Phase coded multi patch, (12cm × 5cm), 3 bits, BW = "*350MHz*", [BALB 09b].

④ Cavity tag (Δf = 2 MHz) (3 × 6cm), 13 bits, BW = 170 MHz, [DEE 10].

⑤ Cavity tag (Δf=1MHz) (3 × 6cm), 16 bits, BW = 170 MHz, [DEE 10].

⑥ "C" tag (Δf=50MHz) (1.5 × 2cm), 9 bits, BW=2200 MHz, [VEN 11a].

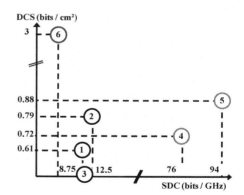

Figure 3.18. *Density of coding per surface unit DCS as a function of the spectral density coding SDC for different chipless tag designs. The numbered references 4, 5 and 6 correspond to designs that have been developed with the aim of improving the surface or spectral density coding and will be the subject of a detailed description in the following chapters. For a color version of this figure, see www.iste.co.uk/vena/ chipless.zip*

We can note that several trends emerge. A DCS-based approach [VEN 11a] will allow us to obtain a very large density of coding per surface unit. Other approaches lead us to favor SDC [DEE 10], particularly by using systems that exhibit very precise frequency resolutions. Finally, other designs allow us to make a compromise [JAL 05b, PRE 09c] between the surface and the occupied bandwidth.

3.9. Conclusion

This chapter has allowed us to introduce the different possible coding techniques in a chipless RFID system. To the extent that none of the studies treat the "coding concept" aspect for the chipless technologies, it is particularly important to conceptualize this notion. This allows us to classify all studies carried out and, on the other hand, to compare them. For this, we have adapted the digital modulation principles borrowed from the conventional telecommunication systems into the chipless RFID field, whose coding is performed as a function of time or frequency. The key elements which allow us to improve the coding efficiency have also been identified. In fact, the improvement of the coding efficiency is an essential aspect in chipless technology, where reduced-size chipless tags with a large storage capacity are sought. Some graphic representations based on constellation diagrams have been proposed for the purpose of better understanding the coding efficiency and the detection robustness of a chipless RFID system. Finally, performance criteria that allow us to link the three essential aspects of chipless tags (storage capacity, surface and required bandwidth) have been introduced, in order to compare the different designs.

Here, we have discussed the general designs of information coding in chipless technology. It now remains to see how it is possible, in practice, to design tags by implementing these principles. Chapter 4 will thus present the different chipless tag designs based on the most effective coding approaches. The guideline of different designs proposed in Chapter 4 is primarily to increase the tag coding capacity by different means. In fact, in order to be able to compete with barcodes, the first step is to obtain a coding capacity at least equivalent to that of EAN 13, i.e. 43 bits. Coding techniques discussed in this chapter will, therefore, help us achieve this ambitious goal in light of the latest developments in this field.

Design of Chipless RFID Tags

4.1. Introduction

In this chapter, we present the key element of a chipless RFID system, in order to learn more information on the tag design. The performance that can be achieved in terms of reading robustness and detection range is directly related to the tag geometry, and to the way in which the information is coded. According to the intrinsic performance of tags (for example, the impact of such geometric quantities on the tag signature, the quality factors obtained, the independence – the non-coupled character of resonators, etc.), it will be possible to use a more or less effective coding method (Chapter 3), which in the end will determine the coding capacity of the tag, as well as its detection robustness (for example, its reading distance).

The design technique presented in this chapter is original. It is based on the idea of using resonant scatterers. Thus, a tag will be composed of a certain number of these elements depending on the available surface and the desired amount of information. This approach is radically different than the one where the tag is composed of antenna(s), transmission line(s) and a circuit part for information coding. As we will see below, there are many benefits with this method. That is why, it has overcome all other approaches. In fact, currently, almost all designed chipless tags are based on this principle. This design method is the *RF encoding particle* (REP) approach. Additional information on the history of this approach (its differences compared to other methods) is provided in [PER 14].

We begin this chapter with a modeling of the chipless identification problem through the example of the use of a basic resonator. We will continue by defining the different inherent performance criteria in a tag design, such as the value of the radar cross-section (RCS) and the resonator selectivity. Next, we will present a comparative study carried out on different basic geometries which can be used in the manufacturing of a chipless tag. Finally, we will present different devices that allow us to address the problem of the coding capacity increase.

4.2. Classification of chipless technologies

4.2.1. *"Temporal" and "frequency" chipless tags*

In this book, we focus on the development of chipless RFID tags whose information is coded in the frequency domain. In fact, the temporal coding techniques [MAN 09, VEM 07, ZHA 06, ZHE 08] based on the use of a low-permittivity substrate are still very far from achieving the surface density coding of surface acoustic wave (SAW) tags [HAR 02] or frequency chipless tags [PRE 09a, PRE 09b]. The reason is inherent in the coding principle used. In fact, these tags code the information in relation to the position of a pulse over time or to its presence. To sufficiently separate two temporal positions in such a way that the reflected signals do not overlap, it is necessary to add delay lines between each discontinuity. The smaller delay to be created is a function of the transmitted pulse width. Thus, for a pulse width of 1 ns, a delay greater than 1 ns must be created to separate two reflections. To increase the number of temporal positions, it is, therefore, necessary to either increase the number of delay line sections or decrease the pulse width. In both cases, we encounter the following problems:

– the increase in the number of line sections leads to a larger tag surface. In addition, at each new discontinuity (at the beginning of the reflection of a part of the incident wave), the signal amplitude decreases, which makes it twice us difficult to increase the line length;

– a very short pulse provides a very spreadout power spectral density in the frequency spectrum. However, it is not easy to achieve in practice the decrease in the pulse width. In addition, it greatly increases the cost of the chipless tag reading system. The maximum usable bandwidth without license specified by the Federal Communications Commission (FCC) for UWB communications is spread out between 3.1 and 10.6 GHz. This defines a minimum pulse width of approximately a few tens of picoseconds. In addition, a filter in the pulse generator output of the reader must be used, to comply with the mask imposed by the standards on UWB communications.

The research carried out in this field was also intended to decrease the number of line sections and their sizes, by using a quadrature phase-shift keying (QPSK) coding technique (thus improving the coding efficiency) with the use of composite right/left-handed (CRLH) lines, which allow us to convey slow waves [MAN 09]. Unfortunately, this architecture is still far too complex, and very difficult to achieve. In addition, the necessary dimensions are large (20 cm long), to code just 6 bits [MAN 09]. It is, therefore, completely unrealistic from a practical and a non-competitive point of view (on the economic plan where the idea remains) to be able to obtain tags with a unit cost of less than a euro cent.

For these reasons, in order to overcome the identified obstacles concerning the implementation of chipless technology, it is necessary to move toward the design of chipless tags, which are coded in the frequency domain. We also note that in contrast to most of the temporal tag conceptions, the possibility to create the entire tag via low-cost processes, such as printing, is quite possible with the proposed designs which are based on a frequency coding [JAL 05b, MCV 06, PRE 09c]. "Frequency" tags have a much better potential for development than "temporal" tags, in any case with regard to the information coding. We will see below the different implemented approaches, which allow us to overcome the problems which arise during the update on the current chipless technology, which were mentioned in Chapter 2.

4.2.2. Circuit approach or use of resonant scatterers

We can group the tags using a frequency coding approach in two large families [PER 14]. The first family, which was introduced by Preradovic *et al.* [PRE 09b], uses a receiving antenna connected to a multi-band notch-filter (see Figure 4.1). The output of this filter is connected to a transmitting antenna which is in cross polarization in relation to the receiving antenna, to isolate the transmission/reception signals. In this case, we have three distinct elements, and for each element there is a very precise feature. The transmitting/receiving antennas which are used are ultra-wideband, in order to be able to cover a frequency band that can go from 3.1–10.6 GHz. In fact, as coding is carried out in the frequency domain, the number of bits that can be coded will be directly proportional, in the case of an absence/presence coding, at the usable bandwidth.

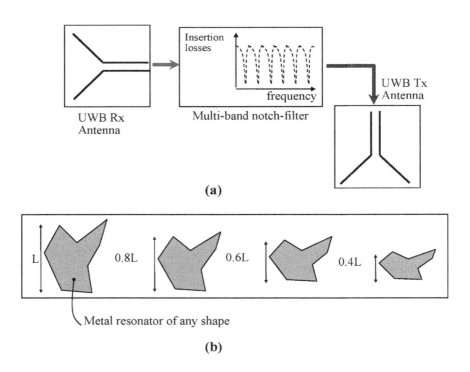

(a)

(b)

Figure 4.1. *a) Approach with two antennas and a filter passive circuit; b) tags in resonant scatterers*

The second family uses the REP approach that differs to the extent that it allows us to obtain more compact structures. In fact, as we have seen in Chapter 2, the same type of electromagnetic response can be obtained with a structure using resonator backscattering elements [JAL 05b, MCV 06, VEN 11c, PER 11] which integrate the three above mentioned functions. This means that this compact element (see Figure 4.1(b)) acts as a receiving antenna, a filter and a transmitting antenna. At first sight, we might assimilate these backscattering elements into any radar targets. But to be more precise, we should note that the choice of this basic element must be carried out in such a way that it can generate a chosen electromagnetic signature and not of any kind. Consequently, there must be a direct and bijective link between the ID coding and a tag-specific geometry. "RF barcodes", presented in [JAL 05b] to transpose the optical barcodes in RF, are somehow the precursor tags of those introduced in this book through the REP design approach [PER 14]. These tags were described at the time as dipole antennas in short-circuit. They are nothing more than resonant backscattering elements, and therefore have all of the specificities theorized in REP approach.

They had, however, a low coding capacity (approximately 5 bits compared to 35 bits of the solution with a line and the proposed antennas in [PRE 09b]), but it is still interesting to compare the surface density coding of each of these two prototypes. In fact, the chipless tag size is also an important element that we should take into consideration. The design by Jalaly *et al.* [JAL 05b] thus offers a much better surface density coding with 0.8 bits/cm² compared to 0.6 bits/cm² for the solution of [PRE 09b]. The coding used in both cases is identical. This shows the miniaturization potential of this approach. REP approach, which is presented below, consists also of the "frequency" chipless tag design based on the combining of resonant backscattering elements, as shown in Figure 4.1(b), and providing a very good compromise between the coding capacity and the tag surface.

4.3. Problem modeling: example of a basic resonator

Here we will introduce the REP design, i.e. the chipless tags design based on the use of multiple resonant backscattering elements. For this purpose, we will study the behavior of a reflector which can be seen as an object of any geometry, and which is composed of a conductive surface. Once excited by an electromagnetic wave, it reflects, or diffuses, a part of this wave in a particular and predictable way. This backscattered wave will be the tag signature. The principle here is to create resonance peaks, which are distinguishable from each other, in the chosen frequency spectrum. The phase change around the resonances can also contribute, as we will see below. For this purpose, we must use reflective elements with a very resonant behavior. Thus, we will now discuss resonators, to designate the backscattering items that are described here, i.e. which are characterized by a resonant behavior. From the electromagnetic response of a resonator, it is possible to extract a model that allows us to create a link between the geometry of the reflector and its ID. From a practical point of view, a model enables us to predict the tag electromagnetic signature and thus to quickly generate the corresponding geometries without requiring a complicated simulation. This specific point, even though it is only rarely mentioned in the literature, is of no less importance. We will see below that the model to associate with tags is different depending on whether a ground plane is used or not. It should be noted that the decoding aspect (ID recovery) is another problem that occurs independently of the structure generation mode. It requires algorithms that implement signal processing functions and will be analyzed in Chapter 5.

To begin, we will discuss the basic element on which we focus, to introduce the chipless tag design method. This element is the resonator in a C-shape which, for reasons of simplicity, will be mentioned below as a "C" resonator. This resonator can be seen as a dipole in short-circuit folded in its center, or even half of a rectangular patch in which a slot is present. Generally, the important point is that a backscattering element can always be considered as an antenna loaded by a complex impedance. To compared with the simple dipole which is wideband, the interest of this "C" structure is its ability to generate a very selective resonance peak (without the presence of a ground plane). A resonance is produced when a quarter of the incident signal wavelength is equivalent to the electrical length of the slot formed by the two "C" arms. This allows the reduction of the maximum resonator dimension by a factor of 2 compared to the dipole in short-circuit. As we will see below, an analytical model can be extracted to connect the slot length at the resonance frequency. In addition, the combination of several resonators is possible in very reduced surfaces, as the coupling between the elements is relatively limited. This coupling aspect is very important in chipless technology to the extent that a very weak coupling between the different resonators will allow us to obtain the bijective link between the resonance frequencies of a part (and thus of the code) and the geometry of the element [PER 14].

4.3.1. *Backscattering mechanisms*

Here, we will see in detail the mechanisms that allow us to excite a conductive object in order to extract its electromagnetic signature. A query signal in the form of an ultra-short pulse, and therefore very wideband, is transmitted by the reader. For reasons of efficiency, the useful frequential components of this signal must be spread out over the entire frequency spectrum which contains the electromagnetic response of the tag that is used to code its ID. A method based on the sending of a narrowband signal which will be able to scan the chosen frequency band can also be used to detect the different frequential components of the tag, as presented in Figure 4.2.

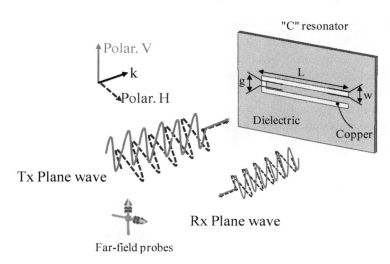

Figure 4.2. *Excitation of a resonator by a plane wave. The recovery of the information of the backscattered field by the tag component by component can be performed using the probes on electromagnetic simulation softwares. The variable parameters in the "C" resonator are the slot length L and the gap g. The width of the metal strip w here is constant*

When the incident wave reaches the tag, it creates currents on the structure as a function of the intensity and orientation of the incident electromagnetic field (polarization) and, of course, as a function of the shape of the conductive element. A part of the incident energy on the tag is retransmitted directly in the entire space, even in privileged regions. This first mechanism of reradiation is structural or specular and depends only on the apparent metal surface [KAR 10]. Another part of the received energy is stored by the structure and rediffused in the entire space according to a process shifted in time, which will depend on the quality factor of the structure. This second phenomenon is called the antenna mode, to refer to the case of an antenna connected to a load [HAR 63, GRE 63]. This reradiation mechanism is shown in Figure 4.3 in the case of a backscattering element which can be assimilated into an antenna connected to a complex load. In a chipless tag based on the combination of resonators, we will seek to modify this antenna mode that is quite resonant and spread out in time, as shown in Figure 4.3(b). As a result, there is a very selective resonance peak, as shown in Figure 4.3(c). Regarding the structural mode, it is inherent in any reflective object and it presents quite a wideband behavior (like the wave reflected by a metal plate of considerable dimensions compared to the wavelength) which can hardly be exploited to code information. In fact, this structural mode is directly related to the object on which the tag is installed. This object is *a priori* unknown. The electromagnetic responses of the tags are, therefore, the integration of structural mode, whose reflected level is almost constant as a function of frequency, with an antenna mode which when superimposed at the structural mode will provide a signature by revealing a peak (additions of phase signals) or a dip (additions of signals against phase and equi-amplitude).

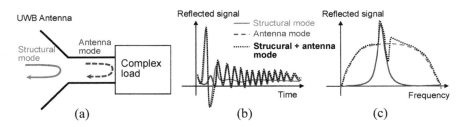

Figure 4.3. *a) Illustration of backscattering mode of a reflective element that can be assimilated to an antenna connected to a complex load, in the form of a structural mode and an antenna mode. b) Impulse response of the reflector illustrating the part of the structural mode and the part of the antenna mode on the total signal. c) Frequency response of the reflector for the total signal, as well as for the two modes taken separately. The addition of the two modes showed a peak and a dip of interference. For a color version of this figure, see www.iste.co.uk/vena/chipless.zip*

To analyze the antenna mode, we can search the current paths in the excited conductive structure. These current paths enable us to link the resonance frequency (through the appearance of a peak in the electromagnetic response) to the resonance mode of the structure and therefore to the geometry. Thus, in the case of a dipole in short-circuit excited by a plane wave in vertical polarization, the first resonance mode takes place when the guided half-wavelength corresponds to the physical length of the dipole (see Figure 4.4). In fact, in this case, we observe a minimum current at the dipole ends and a maximum current at the dipole center (see Figure 4.4(b)). The first higher order mode must also meet the boundary conditions which are imposed by the dipole ends, which means that the current at the ends will always be zero regardless of the resonance mode.

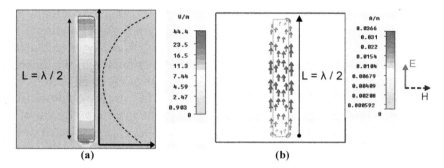

Figure 4.4. *Dipole excited by a plane wave in vertical polarization at its resonance frequency; a) Electric field and b) current density. The dipole length L is close to 50 mm and its resonance frequency is 2 GHz. For a color version of this figure, see www.iste.co.uk/vena/chipless.zip*

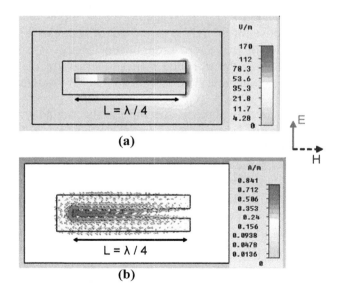

Figure 4.5. *"C" resonator excited by a plane wave in vertical polarization at its resonance frequency: a) Electric field and b) current density. The slot length L is 25 mm, generating a resonance at 2 GHz. For a color version of this figure, see www.iste.co.uk/vena/chipless.zip*

Let us return to the "C" resonator. If we observe in the same way the distribution of its electrical field, Figure 4.5(a), and the current density, Figure 4.5(b), at the resonance, when it is excited by a plane wave in vertical polarization, we note that compared to the largest dimension of the structure, the resonance type is that of a quarter-wave. In fact, boundary conditions are different. On one side of the structure, we observe a maximum current and on the other side we observe a minimum current. We may note that this result could be directly deduced from the current distribution of the dipole in short-circuit, by refolding its ends. The first resonance mode is, therefore, in $\lambda/4$, and the first higher order mode is in $3\lambda/4$. This structure allows us to miniaturize, at a factor greater than or equal to 2, the maximum length of the resonator in relation to the dipole. If we look at the electric field, Figure 4.5(a), we can see that it is concentrated between the two "C" arms. This increases considerably the capacitive effect of this structure. The quality factor of the resonator is proportional to the capacity value. Therefore, frequency selectivity is improved.

Now, if we compare Figure 4.4 with the frequency responses of these two basic elements, which are excited according to the vertical polarization, we observe on one side a very wideband response for the dipole and on the other side a selective response for the "C" resonator. This behavior is related to the energy conservation which is more pronounced for the "C" resonator. In contrast, the amplitude of the reflected signal is larger for the dipole, as it backscatters the electromagnetic field more than the "C" resonator.

When we add a ground plane at the dipole, the frequency response is modified and shows very selective dips, as we can observe in Figure 4.6(a). In this case, a resonant cavity is formed between the conductive strip and the ground plane in the same way as for a patch antenna [BAL 05]. Large quality factors can, therefore, be achieved. In the same manner as the case of a patch antenna, this quality factor, which is related to the bandwidth, depends on the substrate thickness. Unlike the resonance peaks which are observed in the case of a dipole without a ground plane, we observe here only the dip in the spectrum. This phenomenon is related to an interference between the response of the ground plane, a very wideband of high amplitude and the response of the half wavelength conductive strip, at the resonance. The peak is thus masked by the presence of a large structural mode and it is present on the entire frequency band.

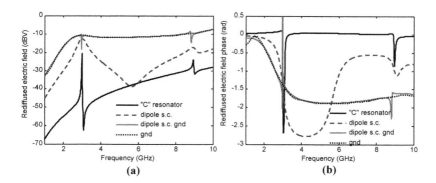

Figure 4.6. *Simulation results obtained with the electromagnetic simulation software computer simulation technology (CST), presenting the responses: a) in amplitude, b) in phase, of the "C" resonator of the dipole in short-circuit with and without a ground plane and of only the ground plane. The parameters of a "C" resonator are: L = 16.175 mm, g = 0.5 mm and w = 1 mm. The settings of the dipole in short-circuit are: L = 28.5 mm, w = 2 mm. The ground plane is 40 mm in height and 30 mm in width. The substrate used is Roger RO4003, with a permittivity of 3.55, a tan δ = 0.0025 and a thickness of 0.8 mm. For a color version of this figure, see www.iste.co.uk/vena/chipless.zip*

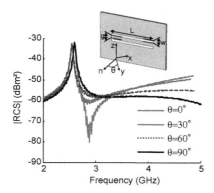

Figure 4.7. *The influence of the incidence angle of the plane wave θ in the azimuth plane on a "C" resonator without a ground plane. The dimensions of the resonator are: L = 18.5 mm, g = 0.5 mm, w = 1 mm. The substrate used is FR-4, with a permittivity of 4.6, a tan δ = 0.025 and a thickness of 0.8 mm. For a color version of this figure, see www.iste.co.uk/vena/chipless.zip*

In Figure 4.7, we can note the influence of the incidence angle on the electromagnetic signature of this resonator. The structural mode is strongly linked to the incidence angle, while this is not the case for the antenna mode which generates the resonance peak [REZ 14]. A detection based on the recognition of the peak positions will, therefore, be more robust with this structure than for a dip detection. However, as we will see below, when we combine several resonators, interference between the resonant modes of resonators is created and produces dips that remain regardless of the incidence angle of the wave arriving on the tag.

4.3.2. *Electromagnetic response modeling*

We can use a transfer function [4.1] to model the resonator. For this purpose, we must analyze in a finer manner the electromagnetic response of the resonator in amplitude and in phase. In the case of a resonator without a ground plane, we can observe around the resonance, for a zero incidence, a peak followed by a dip. The peak can be modeled by a pole of approximately 2. The dip appears when two resonance modes interfere with each other, namely, between the less resonant structural mode and the very resonant antenna mode, when the two signals are in phase opposition. Thus, a backscattered "C" element can be modeled by the addition of two band-pass filters with each having a pole of approximately 2, as shown in Figure 4.8. To generalize this approach in the case of a multi-resonator chipless tag, it is sufficient to add as many terms [4.1] as resonators.

$$T(\omega) = \frac{G_{struct} \cdot e^{-j\phi_{struct}}}{1 + \dfrac{2m_{struct}j\omega}{\omega_{struct}} + \left[\dfrac{j\omega}{\omega_{struct}}\right]^2} + \frac{G_1 \cdot e^{-j\phi_1}}{1 + \dfrac{2m_1 j\omega}{\omega_1} + \left[\dfrac{j\omega}{\omega_1}\right]^2} + \frac{G_2 \cdot e^{-j\phi_2}}{1 + \dfrac{2m_2 j\omega}{\omega_2} + \left[\dfrac{j\omega}{\omega_2}\right]^2} + ... \quad [4.1]$$

In equation [4.1], G_{struct}, G_1 and G_2 represent a gain term which characterizes the response level, respectively, for the structural mode and the first two resonant modes in the case where the tag has two resonators. m_{struct}, m_1 and m_2 represent the damping coefficient and ω_{struct}, ω_2 and ω_3 are resonance angular frequencies of previously mentioned modes. To conclude, each mode has a term that characterizes a phase shift: $e^{-j\phi}{}_{struct}$ for the structural mode, and $e^{-j}{}_{\phi1}$ and $e^{-j}{}_{\phi2}$ for the first two resonant modes. It should be noted that this model does not take into account the higher order modes of resonators (which do not participate in the coding process and are located beyond the bandwidth of the detection system) that intervene at 2–3 times the resonance frequency of the first mode, respectively, for a quarter-wave and a half-wave resonator.

The phase behavior confirms this model. In fact, we can clearly observe a phase jump of $-180°$ at the resonance, followed by a return at $0°$ at the dip level (Figure 4.6(b)). The reason for this dip is the result of a destructive interference between the structural mode and the antenna mode [KAR 10]. In fact, at a precise frequency shortly after the resonance, the antenna mode and the structural mode are in phase opposition and overall of equal amplitude, which causes this dip at the level of the total amplitude of the backscattered field. However, it should be noted that the presence of this dip is strongly linked to the manner in which the tag is excited, particularly at the incidence angle of the plane wave in the azimuth plane. It is, therefore, necessary to balance each term by a constant G_{struct} and G which vary according to the detection configuration [4.1]. The results presented in Figure 4.7 illustrate this behavior. We can note that the presence of the dip is strongly linked to the incidence angle of the plane wave. The structural mode depends on the angle θ, as the equivalent surface facing the plane wave, which is at the origin of the specular reflection of the incident signal, varies, while the resonant mode which generates the peak remains unchanged.

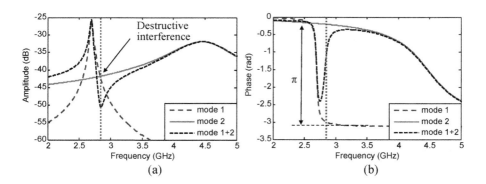

(a)   (b)

Figure 4.8. *Illustration of the interference phenomenon by using the expression [4.1], between the very resonant antenna mode (mode 1) and the structural mode (mode 2) in the case of a resonator without a ground plane. Backscattered signal: a) in amplitude, b) in phase. The parameters of the model with two resonance modes are the following: $f_{struct} = 2.7$ GHz, $f_1 = 4.5$ GHz, $G_{struct} = 0.005$, $G_1 = 0.001$, $m_{struct} = 0.1$ and $m_1 = 0.01$. Modes 1 and 2 are initially in phase $\varphi_{struct} = \varphi_1 = 0$. For a color version of this figure, see www.iste.co.uk/vena/chipless.zip*

For resonators with a ground plane such as a patch dipole in short circuit [JAL 05b], as mentioned previously, we observe only a dip around the resonance. This phenomenon appears because the antenna mode is initially dephased in relation to the structural mode of 90°. This precise value reflects a delay between the direct (or specular) reflection of the structural mode and the reflection of the antenna mode. The structural mode is generated mainly by the ground plane. However, as we have previously explained, the antenna mode is linked to the resonance of the dipole that behaves as a patch antenna. A term of phase shift $e^{-j\varphi}$ for each mode is, therefore, present in the model. In Figure 4.9, we present the behavior in frequency of two modes that interfere, of which one is initially dephased by $\pi/2$ (see Figure 4.9(b)). Thus, we are approaching the behavior of a dipole in short-circuit with a ground plane obtained in a simulation (see Figure 4.6(a)).

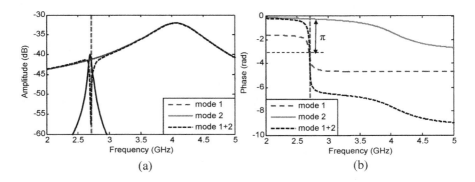

(a)   (b)

Figure 4.9. *Illustration of the interference phenomenon by using the expression [4.1], between the very resonant antenna mode (mode 1) and the structural mode (mode 2) in the case of a resonator with a ground plane: a) in amplitude, b) in phase. The parameters of the model with two resonance modes are the following: f_{struct} = 2.7 GHz, f_1 = 4.1 GHz, G_{struct} = 0.005, G_1 = 0.001, m_{struct} = 0.1 and m_1 = 0.01. Mode 1 is dephased initially by $\pi/2$ in relation to mode 2, therefore φ_{struct} = 0 and φ_1 = $\pi/2$. For a color version of this figure, see www.iste.co.uk/vena/chipless.zip*

In order to validate this model, we have superimposed the spectral signature which is extracted from the simulation in plane wave under CST with the frequency characteristic which is extracted from the model (see Figure 4.10). We note that the model can accurately describe the frequency behavior of a "C" resonator around its resonance frequency and on a frequency range of 2 GHz.

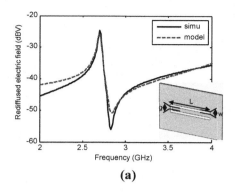
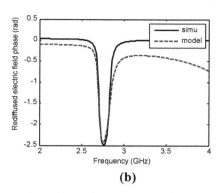

(a)   **(b)**

Figure 4.10. *Electromagnetic response simulated under CST ("simu" curve) and modeled with the relationship [4.1] of "C" resonator around its resonance frequency, a) in amplitude, b) in phase. The parameters of the model with two resonance modes are the following: f_{struct} = 4.5 GHz, f_1 = 2.7 GHz, G_{struct} = 0.005, G_1 = 0.001, m_{struct} = 0.1 and m_1 = 0.01. Modes 1 and 2 are initially in phase φ_{struct} = φ_1 = 0. The parameters of a "C" resonator are the following: L = 18.5 mm, g = 0.5 mm and w = 1 mm*

In the general case of a multi-resonator tag, a circuit model as shown in Figure 4.11(a) can also reproduce the frequency behavior of the "C" structure. It may be configured as a function of the width w of the conductor, of the length L of the device and of the width of its gap g between the two arms.

The structural mode can be modeled by a not very selective series resonant circuit whose resonance frequency is set by the discrete components L_{struct} and C_{struct}, whereas the quality factor is set by the resistance R_{struct}. The overall response level of the structural mode is controlled by the source voltage V_{struct}. The antenna mode interferes with the response of the structural mode and can be modeled by a series LCR circuit which is connected in parallel to the first circuit. The components associated with the first resonant mode are L_1, C_1, R_1 and the source is V_1. This LCR circuit can be very selective as a function of the ability of the resonant element to conserve energy. To model a multi-resonator chipless tag, it is sufficient to add series resonant circuits in parallel. In this model, we have separated the voltage sources, in order to modify the proportion of energy that is caught by the resonant modes and the structural mode. The voltage sources can be dephased between each other to handle the case of resonators without a ground plane or a large phase difference is visible between the structural mode and the other modes. In the case of a "C" resonator, the circuit in Figure 4.11(b), which has a structural mode and a single antenna mode, was simulated with the circuit simulator "Pspice". To find out the rate of the reflected signal as a function of frequency, a current probe has been placed at the radiation resistance level. A comparison between the simulation results of a "C" resonator obtained under CST and the simulation results obtained under Pspice are presented in Figure 4.12. The results obtained on a frequency range of 2 GHz are perfectly consistent and validate this circuit model.

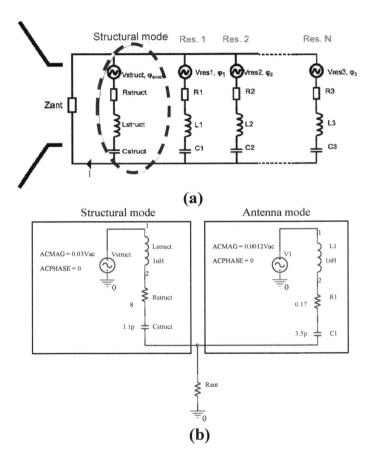

Figure 4.11. *a) Outlined circuit model of a multi-resonator chipless tag, which is excited by a plane wave. b) Simulated circuit model under Pspice of a "C" resonator presented in Figure 4.2, of a resonance frequency 2.7 GHz. A current probe allows us to measure the current in the radiation resistance Rant. For a color version of this figure, see www.iste.co.uk/vena/chipless.zip*

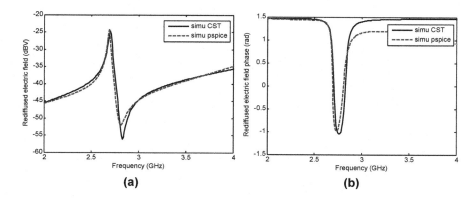

Figure 4.12. *Comparison of the electromagnetic responses simulated under CST Microwave Studio ("simu CST" curve) and under Orcad Pspice ("simu pspice" curve) of a "C" resonator around its resonator frequency, a) in amplitude, b) in phase. The parameters of the circuit with two RLC resonators in parallel are the following: Rstruct = 8 Ω, L_{struct} = 1 nH, C_{struct} = 1.1 pF, $V_{struct}°$ = 0.03 V, φ_{struct} = 0°, R_1 = 0.17 Ω, L_1 = 1 nH, C_1 = 3.5 pF, V_1 = 0.0015 V, φ_1 = 0°. The parameters of a "C" resonator which are presented in Figure 4.2 are the following: L = 18.5 mm, g = 0.5 mm and w = 1 mm*

4.3.3. Radiation pattern

Another important characteristic of a resonator is its radiation pattern or, to be more precise, its reradiation pattern. In this case, it is preferable to refer to the RCS, which is a quantity depending on the polarization of the incident wave (see Figure 4.13) and the direction of observation.

We note that at the resonance, "C" resonator has an isotropic RCS in the azimuth plane (see Figure 4.13(b)), as for a dipole. This is in fact a potentially detectable object regardless of the position of the receiver around the tag. Below, we will see that the tags with a ground plane do not have this advantage. However, chipless tags are intended to be placed on objects, which sometimes prevents any detection by the back side of the tag. In this case, the presence of a ground plane to protect the tag may be beneficial, even necessary. Thus, we note that in the absence of a ground plane, the resonance frequency will change depending on the nature of the object on which we place the tag. This will have a direct impact on the coding, and to overcome this adverse effect, it is essential to implement particular decoding techniques which are described below.

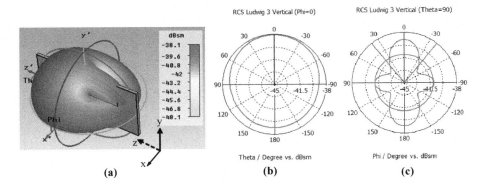

Figure 4.13. *Radar cross-section (RCS) in copolarization that can be interpreted as the reradiation pattern of the "C" cell at its resonance frequency of 2.7 GHz: a) 3D representation; b) H-plane; c) E-plane. The resonator is excited by a plane wave propagating in the X direction, in vertical polarization (following y). The parameters of the resonator are the following: L = 18.5 mm, g = 0.5 mm, w = 1 mm (see Figure 4.2). For a color version of this figure, see www.iste.co.uk/vena/chipless.zip*

4.3.4. *Polarization*

"C" resonator, like a dipole, is an element whose spectral signature depends strongly on the polarization of the incident wave. To excite a dipole, the polarization of the electric field must be aligned with the dipole as in the case of a conventional dipole antenna. Regarding "C" resonator, we must also excite it with a vertical polarization, as shown in Figure 4.2. With a horizontal polarization, the resonance mode does not exist in the expected band of interest. We also obtain conversely a very wideband response. We can also add that contrary to the tag using the approach with two antennas [PRE 09b], a "C" resonator or a dipole are elements that generate a response which can be regarded as having the same polarization as the incident field. However, in order to introduce a more general methodology, we have represented the four possible configurations that allow us to obtain all of the electromagnetic responses characterizing a reflective element (see Figure 4.14). We will see below that it is possible to achieve structures which allow us to promote the creation of a cross-polarized response. The copolarized responses are represented by the terms VV (TX Vertical – RX Vertical) and HH (TX Horizontal – RX Horizontal) and the cross-polarized response by the terms VH and HV.

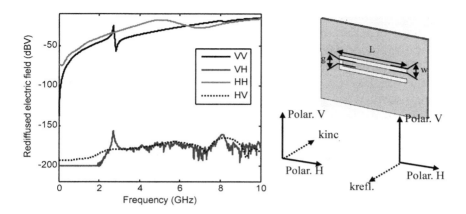

Figure 4.14. *Simulation of the electromagnetic response of a "C" resonator as a function of polarization. VV and VH, respectively, represent the copolarized and cross-polarized responses when the resonator is excited in vertical polarization. HH and HV represent the copolarized and cross-polarized response with an excitation in horizontal polarization. The parameters of the resonator are: L = 18.5 mm, g = 0.5 mm, w = 1 mm. For a color version of this figure, see www.iste.co.uk/vena/chipless.zip*

The horizontal polarization response (HH) presents a very wide band mode which cannot be used to code the information, unlike the vertical polarization response (VV). The cross-polarized responses (VH and HV) are not significant for this "C" resonator. As shown in Figure 4.15, a reradiation pattern can also be obtained via an electromagnetic simulator. Thus, we see that the part of the reflected energy in cross polarization is almost zero in the normal direction at the resonator surface. In contrast, as shown in Figures 4.15(a) and (c), a part of the energy is reflected on the right side of the resonator in cross polarization.

Through the example of a "C" resonator, we have presented the physical principle which is the basis of the communication of a chipless RFID tag. The electromagnetic response of such an object can be modeled via a simple circuit model or a second-order transfer function. As we will see below, it is possible to link the parameters of this model to the tag geometry in the case of a simple structure such as a "C" resonator. This electromagnetic response with a peak followed by a dip is the result of interference between two inherent signals in this type of resonator. We will see below that by combining several resonators in order to create a chipless tag, interference between the different resonance modes and the structural mode will appear. A generalization of the circuit model and the transfer function introduced in this section can be made, in order to model the complete response of a multi-resonator chipless tag, while taking into account its geometry. In the following section, we are going to compare several resonant backscattering elements and define performance criteria that will direct our choice during the chipless RFID tags design.

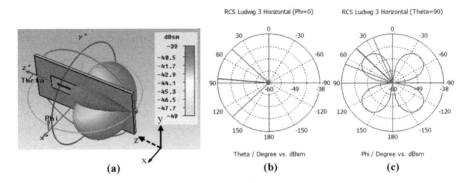

Figure 4.15. *Reradiation pattern (RCS) in cross-polarization of a "C" resonator at its resonance frequency of 2.7 GHz: a) 3D diagram; b) H-plane; c) E-plane. The resonator is excited by a plane wave propagating in the x direction, in vertical polarization (following y). The parameters of the resonator are: L = 18.5 mm, g = 0.5 mm, w = 1 mm (see Figure 4.2). For a color version of this figure, see www.iste. co.uk/vena/chipless.zip*

4.4. Parametric study of basic resonators and performance criteria

We have just explained the physical principles that allow a basic structure such as a "C" resonator to radiate an electromagnetic signature containing specific elements and differentiators. In this section, we are going to compare several elementary resonators according to essential criteria, such as RCS level or spectral selectivity. These criteria will allow us to estimate the performance of a chipless RFID tag that will be created by the combination of several resonators of the same type.

4.4.1. *Determination of performance criteria*

The first essential criterion sought in a resonator is its frequency selectivity. In order to ensure a dense coding in a given bandwidth, it must be able to generate resonances with very narrow bandwidths. This is what will allow us to improve frequency resolution and therefore to considerably increase coding capacity. This is as much valid as the coding principle which is preferable to use is the frequency pulse position modulation (PPM) which was introduced in Chapter 3. Obviously, the selectivity is linked to the quality factor of the resonator which must be the largest possible.

The second criterion, which is also essential, is the RCS, which will largely define the detection range of the resonator. As we will see below, RCS value intervenes directly in the calculation of the power budget to estimate the reading range of the tag.

The reradiation pattern at the resonance is an element which should be taken into account, to verify that the maximum amount of energy is well reflected in a normal direction at the tag surface. In fact, in most cases, detection systems are composed of a single transmitting/receiving antenna or of two separate antennas but which are close to one another (bistatic configuration). Thus, we understand the need to use resonators that meet this criterion.

The surface which is occupied by the resonator, its size and its maximum length in relation to the wavelength must be as small as possible, to ensure a high-surface density coding. Thus, in the worst case, the largest dimension of the resonator will be equal to the half-wavelength (for example, the dipole in short-circuit).

Finally, another element which we should take into account is the position of the first higher order mode. In fact, it will define the upper limit of the usable bandwidth. In order to optimize the potentially permitted bandwidth (ultra-wideband (UWB) band), it will be expected that this upper limit is rejected beyond 10 GHz or at the highest possible value if this value cannot be reached. In the best case, the frequency of the first higher order mode is equal to 2 or 3 times the frequency of the fundamental mode, respectively, for a half-wave or a quarter-wave resonator.

4.4.2. *Comparison of resonators*

By using the same approach as the one that has been implemented through the example of a "C" resonator, we can extract the information corresponding to the performance criteria which were previously defined. Thus, in Table 4.1, we have gathered these different criteria for the main structures with a resonant behavior. In order to compare the performance of each structure, the substrate used is the same in all cases (thickness, dielectric constant and losses). All elements have been simulated under CST Microwave Studio with the same approach and an optimization of each element has been carried out for the same resonance frequency (3 GHz).

Type	Image	Dimensions (mm)	BW -3dB	Maximum dimension - multiple of wavelength	RCS (dBm²)	Frequency of the first higher order mode
Dipole in short-circuit		L = 34 w = 0.5	452 MHz	$\lambda/2$	−21.35	8.94 GHz
Simple circ. SRR		R = 5.4 g = 1.6 w = 0.5	72 MHz	$\lambda/6.3$	−25	8.95 GHz
Simple rect. SRR		L = 9.14 g = 1.6 w = 0.5	68 MHz	$\lambda/7.44$	−24.5	8.66 GHz
"S" resonator		La = 10 Lo = 10 w = 0.5	84 MHz	$\lambda/6.8$	−21.9	>10 GHz

"C" resonator cell (L = g)		L = 12.1 w = 0.25	120 MHz	$\lambda/8.5$	−22.9	8.32 GHz
"C" resonator		L = 16 w = 1 g = 0.5	8 MHz	$\lambda/4$	−31.4	8.95 GHz
"Z" curve at the order of 1		L = 12.5 w = 0.5	149 MHz	$\lambda/5.44$	−21.61	>10 GHz
"Z" curve at the order of 2		L = 9.24 W = 0.5	48 MHz	$\lambda/7.35$	−22.76	>10 GHz
Hilbert structure - 2nd order		L = 8.92 w = 0.25	48.6 MHz	$\lambda/7.63$	−23.55	7.73 GHz
Hilbert structure - 3rd order		L = 6.5 mm w = 0.25 mm	18.5 MHz	$\lambda/10.5$	−27.3	7.7 GHz
"C" resonator + dipole		L = 13.3 mm, H = 15 mm, g = 0.5 mm	32 MHz	$\lambda/6.67$	−22.5	6.73 GHz

Table 4.1. *Comparison of the performance of different resonators without a ground plane. The substrate used is RO4003 (with a permittivity of 3.55 and a tan δ = 0.0025), the substrate thickness is 1.6 mm. To characterize the miniaturization aspect of each resonator, the maximum dimension is provided as a multiple of the wavelength λ at the resonance frequency (3 GHz)*

The resonators presented in Table 4.1 have the advantage of operating with a single conductive layer, i.e. without a ground plane. This makes them compatible with low-cost manufacturing processes such as pattern printing techniques directly on the object to identify. The lower bandwidths that can be achieved are approximately 8 MHz. To significantly miniaturize the patterns, Hilbert filling curves or "Z" filling curves can be used. To this, we add the fact that by increasing their order, we tend to make them more selective at the expense of an RCS decrease. A compromise must be made between selectivity, structure size and their RCS. "C" resonator is a basic element which presents a good compromise: it is compact, it presents a good selectivity and it has a higher order mode at almost 3 times its main resonance frequency. In addition, by altering only its gap g, we can increase its selectivity. In fact, as we will see below, another advantage of a "C" resonator is that it can be placed in a vertical arrangement without having too much coupling between the different elements. Finally, in the case where the gap is small, its resonance frequency can be modified simply by adjusting the slot length. It is an undeniable advantage that greatly facilitates the design phase and subsequently the possible reconfiguration of the tag.

Now, we can compare different resonant structures using a ground plane. These structures can be used to promote the selectivity of resonance peaks, as well as to isolate the tag of the object to identify. In Table 4.2, we have gathered different backscattering elements with a ground plane. Here, we are referring to ΔRCS rather than to RCS, to qualify the dip amplitude which is generated by the presence of a resonator in relation to the reflection level of a ground plane without a resonator (see section 4.3.2). RCS level without a resonator is linked to the size and geometry of the ground plane. Therefore, we cannot compare the performance of resonators with a ground plane on the simple criterion of the RCS level. Thus, we can observe that it is very interesting, in this case, to use the dipole in short-circuit, as the dip of interference generated is very selective. It is also the basic element of a chipless tag by Jalaly *et al.* [JAL 05b]. The circle is also selective and also presents a polarization independent response. This makes its use attractive for the creation of chipless tags that can be detected regardless of their orientation. This is what is observed in [VEN 12c]. This study will be analyzed in the following section.

Type	Image	Dimensions (mm)	BW -3dB	Maximum dimension - multiple of wavelength	ARCS	Frequency of the first higher order mode
Dipole in a narrow short-circuit		L = 28 w = 2	7.6 MHz	λ/2	8 dB	8.97 GHz
Dipole in a large short-circuit		L = 26 w = 10	28 MHz	λ/2	5.5 dB	9 GHz
Circular ring.		R = 9.5 w = 0.5	14.5 MHz	λ /2.95	5 dB	9 GHz

Table 4.2. *Comparison of the performance of resonators with a ground plane. The substrate used is Roger RO4003 (with a permittivity of 3.55 and a tan δ = 0.0025), the substrate thickness is 0.8 mm. To characterize the miniaturization aspect of each resonator, the maximum dimension is provided as a multiple of the wavelength λ at the resonance frequency (3 GHz)*

4.5. Combination of several resonators and optimization method

It is now important to define the design rules which we have to use and the optimization methods which allow us to converge toward the desired electromagnetic response. Once the resonator has been chosen and optimized in order to converge toward an RCS level and a required selectivity, we must primarily establish the link between the differentiating characteristic in the spectrum (the presence of a peak, for example, in this case, the important parameter is the resonance frequency of the structure) and one or several geometric parameters of the structure. This physical parameter can be, depending on the case, a slot length, a gap width or simply a magnification/reduction factor of the entire structure. Thus, in the case of a "C" resonator with a gap of 0.5 mm and a conductor width $w = 1$ mm, we can obtain the curve of Figure 4.16(a) by varying the length L between each simulation. From this characterization, we can deduce the dimensions of the resonators that will create peaks or dips (the dips are also related to the resonance frequency) at specific frequencies. For example, let us look at the case of a tag with 5 "C" resonators of a gap $g = 0.5$ mm, which is presented in Figure 4.17. We want to create five distributed peaks between 2 and 3 GHz. In Table 4.3, we find the slot lengths L_1–L_5 of the five resonators (see Figure 4.17(b)). These values are deduced from an extrapolation of a parametric study carried out for a unique "C" resonator (Figure 4.16(a)).

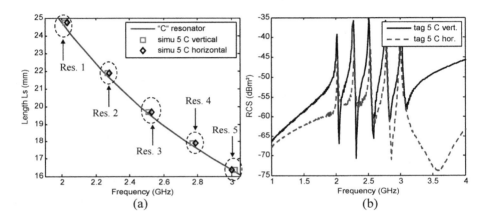

Figure 4.16. a) Link between the value of the resonance frequency and the slot length L for a "C" resonator (w = 1 mm, g = 0.5 mm), and for the 5 "C" resonators arranged in a horizontal (Figure 4.17(a)) and vertical (Figure 4.17(b)) configuration. The squares (respectively, the diamonds) represent the values of the frequency obtained for lengths L_1 - L_5 which are extracted from the parametric study carried out on the unique "C" resonator (char. "C" resonator) in the case of a tag in vertical topology (respectively, horizontal). b) The simulation result of a tag with 5 "C" resonators in horizontal and vertical topology. For a color version of this figure, see www.iste.co.uk/vena/chipless.zip

After a first simulation, we can observe the shifts in relation to the desired frequencies (the gaps between the diamonds – squares is the curve with the results of the parametric study on a resonator). It should be noted that these shifts are strongly linked to the arrangement of the resonators (see Figure 4.17) and their spacing. In fact, the coupling between the different resonators impacts resonance frequencies. In addition, it acts differently depending on the configuration used (see Figures 4.16(a) and (b)). The values of resonance frequencies obtained in a simulation under CST are gathered in Table 4.3. We can note a maximum shift of resonance frequencies of 1% for the vertical topology and 1.5% for the horizontal topology.

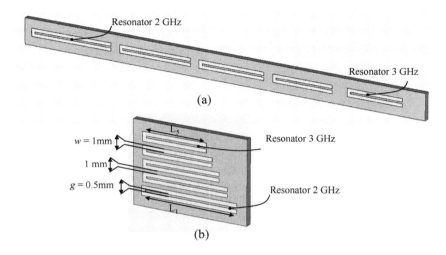

(a)

(b)

Figure 4.17. *Tag with 5 "C" resonators: a) tag with a horizontal arrangement and b) tag with a vertical arrangement*

Topol.	Theor. lengths (mm)					Isol. resonator frequency					Simul. frequencies									
	L1	L2	L3	L4	L5	F1	F2	F3	F4	F5	F1	ΔF1	F2	ΔF2	F3	ΔF3	F4	ΔF4	F5	ΔF5
Vert.	24.8	21.9	19.7	17.9	16.4	2	2.3	2.5	2.8	3	2.02	1%	2.27	0.9%	2.51	0.4%	2.77	0.7%	3.02	0.7%
Hor.											2.03	1.5%	2.28	1.3%	2.53	1.2%	2.79	1.5%	3	0%

Table 4.3. *Resonator dimensions and associated resonance frequencies. The lengths $L_1 - L_5$ and the theoretical frequencies $F_1 - F_5$ are extracted from the parametric study on the "C" resonator, enabling the linking of the resonance frequency to the slot length L. The frequencies obtained in simulations under CST for vertical and horizontal topologies (Figure 4.15) are also provided*

Thus, we can see that it is difficult to predict in advance the coupling between the modes. In contrast, an optimization sequence on the slot length parameters can be implemented to find the characteristic linking the resonance frequency to the slot length for a given tag configuration (i.e. with the couplings). In this case, we have to launch several successive simulations of the complete tag by slightly varying the length of all resonators between each simulation. The extrapolation of all frequency points obtained allows us then to trace an average characteristic, which is truer to the behavior of the complete tag. A polynomial of the order N [4.2] and of a coefficient a_k can in most cases approach precisely this characteristic, by making the link between the resonance frequency f and the length L of an element.

$$f = \sum_{k=0}^{N} a_k \cdot L^k \qquad\qquad [4.2]$$

It is very important to obtain a precise model which allows us to link the resonance frequency to a slot length or a physical parameter. In fact, this enables us to predict the electromagnetic signature of a particular tag to generate the desired ID. This optimization allows us to obtain an average characteristic which must be sufficiently precise to generate the desired resonance frequencies. The residual error related to very slight coupling variations from one configuration to another must be less than half of the chosen frequency resolution for the coding, in order to ensure a reliable decoding. Finally, we can have a model linking a code to the lengths of each resonator. Let us examine the case of a PPM coding transposed in the frequency domain. To make the link between the shift of resonance frequency F_0 of a peak in relation to the initial value f_{i0} and the value of a coding digit C_i, the relationship [4.3] can be used. C_i is an integer, the "low hook", which indicates to take the integer part of the subtraction. Frequency resolution Δf is limited by the resonance selectivity and the measurement system.

$$C_i = \left\lfloor \frac{f_i - f_{i0}}{\Delta f} \right\rfloor \qquad\qquad [4.3]$$

The construction of the complete code, i.e. the tag ID, is based on the association of several digits $C_0 - C_{N-1}$ weighted by a weight which depends on their base and their rank in the code. In fact, in a general case, we can associate a code to each resonator which generates a number of different bits. In this case, equation [4.4] can be used to find the ID. But in most cases, the number of possible combinations for each digit is identical, which allows us to reformulate equation [4.4] in [4.5] in this particular case. In these relations, N is the number of digits and b_k is the basis associated with the digit C_i when $k = i$. For equation [4.5], the base is the same for all digits ($i = 0$ at $i = N\text{-}1$) and it is the term b.

$$ID = C_0 + \sum_{i=1}^{N-2} C_i \prod_{k=1}^{i} b_k \qquad [4.4]$$

$$ID = \sum_{i=0}^{N-1} C_i b^i \qquad [4.5]$$

In the case where an analytic expression allows us to make the link between a resonance frequency and a geometrical length, we can insert [4.3] into [4.2] to obtain [4.6]. Finally, we obtain a direct link between a digit and a physical length, the two associated with a same resonator. In this equation, L_0 represents the basic length, in order to obtain the initial resonance frequency for which we can associate the value 0.

$$C_i = \frac{\sum_{k=0}^{N} a_k \cdot \left(L^k - L_0{}^k \right)}{\Delta f} \qquad [4.6]$$

For the design stage, it seems more appropriate to obtain an inverse relationship compared to [4.6], in order to obtain as output the parameter value L for a given C_i code as input. We observe here that the equations are complicated when the order of the polynomial increases.

4.5.1. *Conclusion*

We have established the basic designs for the generation of a structure based on the combination of identical resonators. We have analyzed the method in which we can obtain a precise model allowing us to link a resonance frequency to a physical parameter with the use of two design stages:

– a first step is to consider the resonator alone, in order to obtain a first characteristic which will provide a more or less precise value for each resonator which constitutes the complete tag;

– a second stage, known as optimization, will help us refine the frequency characteristic/physical parameter(s) by taking into account the coupling between the resonators that depends on their positioning and the spacing between them.

To conclude this section, we have constructed a model which allows us to link a code to a physical length. This model can be used at a later stage in the tag creation phase, to design a particular tag from the chosen ID.

At the remaining of this chapter, we present in detail the different chipless tag designs based on the REP design principle which was introduced in this section.

4.6. Design of tags without a ground plane

4.6.1. *Presentation of design no. 1: double "C" tag*

The first issue that we will discuss in this book concerns the chipless tag miniaturization, as well as the increase in the surface density coding. For this, we can start from the structure presented in Figure 4.18. This structure offers a very good selectivity and requires a reduced surface. The design developed in [VEN 11a] and [PER 11] is potentially very low-cost. In fact, we will show that by taking into account the materials and production techniques, the tag cost would be comparable to that of a barcode label. In fact, it requires only a single conductive layer, which makes it compatible with printing processes directly on the product.

4.6.1.1. *Tag description*

The tag is based on the PPM frequency coding principle. The presence of resonances at the predefined frequencies allows us to code a particular ID, as we have seen in Chapter 3. The structure chosen here is similar to a short-circuited coplanar line on one side and an open line on the other side (see Figure 4.18(a)). This configuration enables an improvement in terms of miniaturization to the extent that resonance appears for a slot length in $\lambda/4$ with a large quality factor. Although the response level is lower than for other resonators with an RCS of -28 dBm^2, by using an FR-4 substrate, this elementary structure provides a certain advantage of miniaturization and coding density. The tag proposed in Figure 4.18(a) is composed of four resonators (1, 2, 3 and 4) independent of each other, apart from No. 3 which acts as an isolation between on the one hand the resonators 1, 2, and on the other hand the resonator 4. In fact, the resonance frequency of No. 3 varies slightly depending on the slot length of the other resonators. It therefore cannot be used to code information. Contrary to the simple "C" resonator which was previously introduced, the slots have right angles, in order to decrease the maximum dimension necessary to create a resonance in $\lambda/4$.

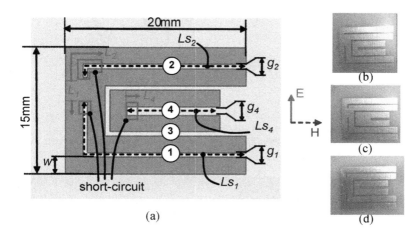

Figure 4.18. *a) Double "C" tag structure, b)–d) images of different configurations of tags 1–3 and created on FR-4. L_{S1}–L_{S4} represent the slot length marked with the numbers 1–4 on the figure. L_1, L_2 and L_4 define only the length of the short-circuits which are placed in the areas demarcated by the red rectangles on the figure. The slot dimensions are: $L_1 = 24.7$ mm, $L_2 = 20.3$ mm, $L_3 = 11.1$ mm. The short-circuit lengths for tags 1–5 are provided in Table 4.3. For a color version of this figure, see www.iste.co.uk/vena/chipless.zip*

Finally, the obtained structure allows us to have three independent resonators which have a spectral response that is exploitable between 2 and 6 GHz, as shown in Figure 4.19, in a very reduced surface, as shown in Figure 4.19. The substrate used is FR-4, with a permittivity of 4.6 and a thickness of 0.8 mm, occupying a surface of 1.5×2 cm^2. The use of a low-cost, low-performance substrate, as is the FR-4, allows us here to show the tolerance of this design at the support. This is a very important point, in order to be able to follow low-cost production techniques. To validate this new chipless tag design, we have carried out five different configurations, whose geometrical parameters are provided in Table 4.4. Photos of the tags 1–3 are shown in Figures 4.18(b)–(d). In accordance with the notations in Figure 4.18(a), the track width is $w = 2$ mm and the gap width is $g_1 = g_2 = g_3 = 0.5$ mm with $g_4 = 1$ mm.

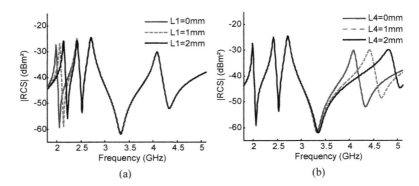

Figure 4.19. *Simulated RCS under CST of the double "C" tag:*
a) For different short-circuit lengths L₁ and b) same thing for L₄.
For a color version of this figure, see www.iste.co.uk/vena/chipless.zip

Configuration	ID	Ls_1	L_1	Ls_2	L_2	Ls_4	L_4
Tag 1	0.2.1	24.7	0	20.3	1	11.1	0.3
Tag 2	5.0.1	24.7	3	20.3	0	11.1	0.3
Tag 3	0.0.15	24.7	0	20.3	0	11.1	3.5
Tag 4	3.0.1	24.7	1.85	20.3	0	11.1	0.3
Tag 5	0.1.1	24.7	0	20.3	0.5	11.1	0.3

Table 4.4. *Dimensions of tags created on an FR-4 substrate (with a permittivity of 4.6, a tan δ = 0.025 and a thickness of 0.8 mm). The values are provided in mm. The expected ID is provided for each configuration. The copper track width for each resonator is w = 2 mm, and the gaps have the following values: g1 = g2 = g3 = g4 = 0.5 mm*

Figure 4.19 presents the frequency variation of the first mode and the fourth resonance mode for different slot lengths, respectively, marked as L_1 and L_4 (see Figure 4.18(a)). These simulation results confirm the independence of these modes regarding the variation of the neighboring modes. To vary the resonance frequency, two techniques are possible. The first technique consists of simply varying the resonator length. The second technique, which is used here, refers to the idea of placing a short-circuit element inside the slot and of varying its length. The interest of this technique is the configurable aspect that it provides. Thus, a batch of blank tags can be produced in large amounts, and then in a second phase, by adding a short-circuit element, each tag can be customized. The short-circuits are represented by the red rectangles in Figure 4.18(a).

On the coding side, several solutions are available to us. It is possible to have a conventional coding which is based solely on the presence/absence of lines to code a bit to 1 or 0. This type of coding is robust, but it has many requirements in terms of surface, as 1 bit is equivalent to 1 resonator. For this reason, the coding used here is based on the PPM principle transposed in the frequency domain and it allows us to combine multiple bits at 1 resonator.

4.6.1.2. *Performance and results achieved*

The measurement of the chipless tag signature has been carried out with the measuring bench described in Figure 4.20. A bistatic radar configuration has been used and a frequency approach using a vector network analyzer (VNA) has been implemented. The idea was to measure the parameter S_{21}, which will provide information on the relation between the wave sent by the transmitting antenna and the wave received by the receiving antenna which is connected to another port of VNA. By using a calibration based on the additional measurement of an empty environment and of a reference whose electromagnetic signature is known, we can trace the RCS which characterizes the tag independently of its detection distance. This calibration procedure will be described in Chapter 5.

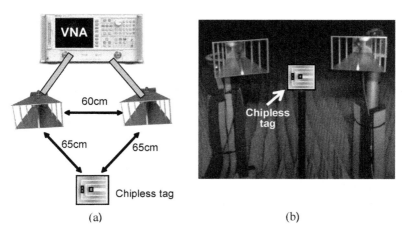

Figure 4.20. *a) Configuration of a bistatic radar measuring bench; b) Photo of the measuring bench of a chipless tag in an anechoic chamber*

For this measurement, the two antennas used are broadband horns whose gain changes between 10 and 12 dBi between 1.5 and 6 GHz. The chipless tag to be measured is placed at 65 cm of each antenna and the antennas are separated by 60 cm. The output power of VNA is 0 dBm. An initial measurement is carried out without a tag. Subsequently, all measurements in the presence of a tag are subtracted from the record of the initial measurement. This process allows us to overcome, among others, the coupling between the two antennas. We will further analyze the configuration of the measuring bench in Chapter 5, which will also address a measuring bench in the temporal domain.

Measurements have been performed on the five tags with variable short-circuit configurations (see Table 4.4) and whose basic structure is provided in Figure 4.18(a). Figures 4.21 and 4.22 present the measurement results of RCS performed with the frequency measuring bench described in Figure 4.19(a). They are compared with the simulation results in a plane wave, performed with CST. We note a very good consistency between the measurements and the simulations. In fact, the four resonance peaks related to four modes are clearly visible for the three tags (Figures 4.21 and 4.22(a)). Their frequencies correspond precisely to those expected by the simulation. Figure 4.22(b) presents the resonance frequency variation of mode No. 2 for different short-circuit configurations L_2, which is 0 mm for tag 3, 0.5 mm for tag 5 and 1 mm for tag 1.

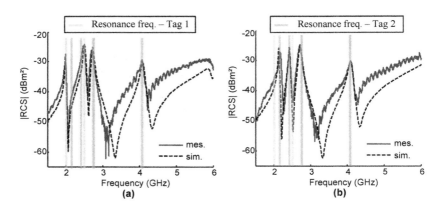

Figure 4.21. *Measurement and simulation results under MWS CST of the double "C" tag (see Figure 4.18). a) Tag 1 and b) tag 2. Between tag 1 and tag 2, the short-circuit L_1 changes from 0 to 3 mm and the short-circuit L_2 changes from 1 to 0 mm. For a color version of this figure, see www.iste.co.uk/vena/chipless.zip*

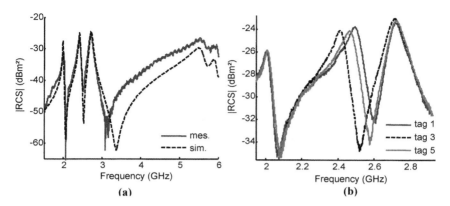

Figure 4.22. *a) Measurement and simulation results for tag 3, b) integration of the measurements performed on tags 1, 3 and 5 around the second resonance mode. The short-circuit lengths L_2 are, respectively, 1, 0 and 0.5 mm for tags 1, 3 and 5. The other dimensions of the tags are provided in Table 4.3. For a color version of this figure, see www.iste.co.uk/vena/chipless.zip*

Configuration	Mode 1 (GHz)		Mode 2 (GHz)		Mode 4 (GHz)	
	Meas.	Simul.	Meas.	Simul.	Meas.	Simul.
Tag 1	1.98	2	2.49	2.5	4.09	4.1
Tag 2	2.22	2.25	2.44	2.4	4.09	4.1
Tag 3	2	2	2.42	2.4	5.53	5.5
Tag 4	2.18	2.15	2.42	2.4	4.1	4.1
Tag 5	2.01	2	2.46	2.45	4.1	4.1

Table 4.5. *Resonance frequencies measured and simulated under CST as a function of the tag configuration. The dimensions of tags 1–5 are provided in Table 4.3*

In Table 4.5, we have gathered the resonance frequencies associated with each resonator, which have been extracted from measured and simulated frequency signatures under CST (see Figures 4.21 and 4.22). The mean error between the resonance frequencies obtained by simulation and those obtained by measurement remains below 30 MHz. Thus, we observe that it possible to accurately predict the behavior of the different tag configurations by simulation. Between tag 1 and tag 2, a variation of 3 mm on the short-circuit length L_1 produces a shift of 220 MHz on mode 1. Similarly, a variation of 1 mm on L_2 introduces a shift of 50 MHz on mode 2. Mode 4 remains logically unmodified since L_4 does not vary. Between tags 2 and 3, a variation of L_4 equal to 3.2 mm introduces a frequency jump of mode 4 close to 1,500 MHz. Undesired frequency shifts of approximately 20 MHz can be observed on mode 1 between tags 1 and 3 and on mode 2 between tags 2 and 3. To compensate for these untimely shifts, we set a minimum frequency resolution of Δf = 50 MHz. From that point and by knowing the variation range of each mode, the coding capacity of this design can be deduced by using the relation [3.11]. Mode 1 may vary in a range of 2–2.4 GHz, mode 2 varies between 2.4 and 2.7 GHz and mode 4 varies between 4 and 5.5 GHz. This provides, respectively, a bandwidth of 400, 300 and 1,500 MHz. Thus, the number of possible combinations for mode 1 is equal to 8, 6 for mode 2, and 30 for mode 4. We obtain a total capacity of 8 × 6 × 30 = 1,440 combinations or 10 bits in 1.5 × 2 cm². This provides a surface density coding of 3.3 bits/cm². It is a coding density by a surface unit of 4–5 times higher than the previously introduced designs [PRE 09b, JAL 05b].

4.6.2. Presentation of design no. 2: "C" tag with 20 elements

The previous design has enabled us to obtain a large surface density coding due to the use of quarter-wavelength resonators and an efficient coding technique. In contrast, we observe very quickly the limitation of such a device when we would like to increase the number of resonators. Unusable modes appear and restrict the bandwidth. In addition, resonators cannot be indefinitely embedded in the frequency band of interest (3.1–10.6 GHz). In the following study [VEN 12a], we have sought to increase the coding capacity by combining several "C" resonators separated from each other by a gap, in order to ensure a decoupling. The coding technique used in this case is simply based on the absence/presence of a peak in the spectrum.

4.6.2.1. Tag description

The resonance frequency of a "C" resonator depends on the length L of the slot and its gap g, while its quality factor is defined in relation to L/g

(see Figure 4.2). In fact, when the two "C" arms are closer, field lines are more dense in such a way that the quality factor is increased. The chipless tag presented in Figure 4.23 is designed with 20 resonators, which allows us to create 20 resonance peaks in the allocated spectrum (the results are presented in Figure 4.24). The coding principle used simply associates 1 bit for each resonance. For 20 resonators, coding capacity is therefore 20 bits, or more than 1 million combinations. To configure the tag, each slot can be completely short-circuited or not (see Figure 4.23) according to the ID to generate. In this way, if the resonator is covered with a conductive material, its resonance frequency is shifted toward higher frequencies located outside the frequency band of detection (around 30 GHz since the height of the filled rectangle is equal to 2.5 mm). In order to show the tag performance, the tag has been created on a common and very versatile substrate, the FR-4. Its permittivity is 4.6 and its thickness 0.8 mm. The surface which is necessary to contain 20 resonators is 70×25 mm². The track width w of "C" resonators is 1 mm and the gap is 0.5 mm. The resonators have been designed so that they can resonate between 2 and 4 GHz. We can thus observe in Figure 4.24 a peak every 100 MHz. In order to minimize the tag size, the resonators are separated only by 1 mm. For this spacing, the coupling effect is present. However, this phenomenon can be taken into account during the design by adjusting the resonator length. Tag 1 has 20 resonators which are free to resonate between 2 and 4 GHz. We observe 20 resonance peaks in Figure 4.24. Resonators 2, 11 and 19 of tag 2 are covered with metal, same as resonators 1 and 4 for tag 3. Thus, we observe in Figure 4.24, some missing resonance peaks which correspond to these last two configurations. In fact, when a resonator is covered with metal, the associated peak disappears and the neighboring resonant mode at the closest frequency has an amplitude increase, and its resonance frequency is very slightly shifted. However, this slight shift of approximately 10–20 MHz is not problematic if we set, as in this case, a frequency resolution of 100 MHz.

4.6.2.2. *Performance and results achieved*

Measurements have been carried out in an anechoic chamber with the same procedure as that described in section 4.6.1.2. Here the tag is placed on a support at a distance of approximately 50 cm of each antenna. We have measured the responses of three tags in amplitude, as well as in phase. As we can observe in Figure 4.25, the amplitude response provides results close to that of the simulation. However, it should be noted that in an environment disordered by multiple echoes and noise, these results can quickly become unusable.

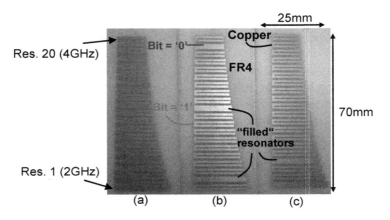

Figure 4.23. *Photos of "C" tags with 20 elements: a) Tag 1; b) Tag 2; c) Tag 3. Each resonator codes a bit. Rather than removing the resonator to code a 0, the approach is to short-circuit the slot of the resonator in question. The basic "C" resonator is described in Figure 4.2. The track width w is set at 1 mm and the gap g is equal to 0.5 mm. The space between two consecutive resonators is 1 mm. The resonator 1, which generates a mode at 2 GHz, has a slot length of L = 23 mm. The resonator 20, which generates a mode at 3.9 GHz, has a slot length of L = 11.5 mm*

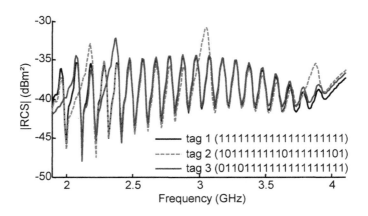

Figure 4.24. *Simulation results under CST for the 3 "C" tags with 20 elements presented in Figure 4.23. For a color version of this figure, see www.iste.co.uk/vena/chipless.zip*

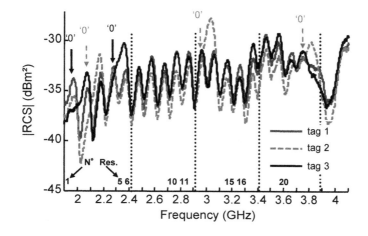

Figure 4.25. *Measurement results in amplitude for the 3 "C" tags with 20 elements presented in Figure 4.23. For a color version of this figure, see www.iste.co.uk/vena/chipless.zip*

From the phase of the reflected signal extracted from the complex parameter RCS (in practice, from the complex signal S_{21} measured with the use of VNA), we can determine the evolution of the group delay as a function of frequency. As a general rule, RCS is a real value which describes the ability of an object to sense a power density, so that it can retransmit it in particular directions. Here, we use an extended version of this definition by adding an imaginary part to describe the phase behavior of this object. As we have explained in Chapter 3, this parameter is robust, as it is not affected in the same way by the reading distance as the amplitude. Thus, we can compare the information provided by the amplitude in Figure 4.25 and by the group delay in Figure 4.26. The peak level in the signal expressed in group delay is generally constant for all resonators. This makes the decoding task simpler and more robust than for a detection from the amplitude. However, in both cases, we can locate the position of "0s" and "1s" without error.

In summary, here, we see that it is possible to code 20 bits in a relatively small surface of 70×25 mm². Thus, we obtain a surface density coding of approximately 1.14 bits/cm². We also see that group delay is a parameter which proves to be efficient for the detection of chipless tags.

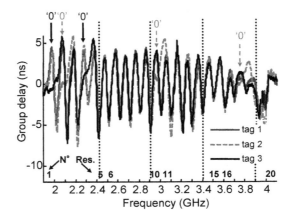

Figure 4.26. *Measurement results in group delay for the 3 "C" tags with 20 elements presented in Figure 4.23. For a color version of this figure, see www.iste.co.uk/vena/chipless.zip*

4.6.3. *Presentation of design no. 3: simple "C" tags at hybrid coding*

The following study focuses on the improvement of the coding capacity by working on the coding capacity [VEN 11c, VEN 11b]. This allows us to limit the number of "C" resonators, thus reducing the tag size, but at the expense of a larger occupied band. A hybrid coding technique has been introduced in order to increase the number of possible states for the same resonator. For this, a coding based on the evolution of the frequency phase [VEN 11b] is combined at a frequency PPM coding [VEN 11c].

4.6.3.1. *Tag description*

The tag which is presented in Figure 4.27 is based on the combination of 5 "C" resonators that are contained in a surface of 2×4 cm². Unlike previous studies where only the slot length L was used to modify the resonance frequency, here we also modify the gap value g between the two "C" arms, in order to modify the interference dip position for each resonance. In fact, at the beginning of this chapter, we saw that an interference dip occurs when the structural mode and the resonant mode are in phase opposition. Therefore, the phase modification will vary the position of this dip. In this way, we can combine a frequency PPM coding with a phase coding.

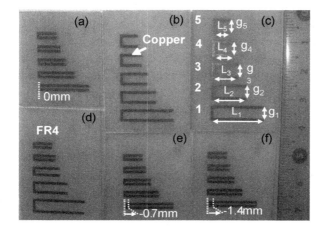

Figure 4.27. *Photo of tags with 5 "C" resonators in hybrid coding:*
a) tag 1; b) tag 2; c) tag 3; d) tag 5; e) tag 6 and f) tag 7. The dimensions
of the tags are provided in Table 4.5. For a color version
of this figure, see www.iste.co.uk/vena/chipless.zip

For this study, eight tags with different configurations have been created, some of which are represented in Figure 4.27. In Figure 4.28, we can see in a simulation, the amplitude and phase evolution as a function of frequency for several gap values *g*.

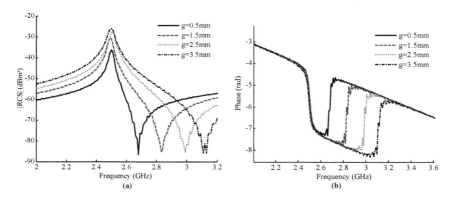

Figure 4.28. *Evolution of the simulated electromagnetic response of a "C" resonator as a function of the value of its gap g, for a length L equal to 18.5 mm and a track width of 1 mm. (a) RCS and (b) phase of the reflected signal for a distance of 100 mm between the tag and the detection probe. For a color version of this figure, see www.iste.co.uk/vena/chipless.zip*

The behavior observed in amplitude and phase corresponds to the model described at the beginning of this chapter with equation [4.1]. The constant decrease observed on the phase is related to the distance between the tag and the field probe placed in the simulation environment under CST. The greater the distance, the more the slope is marked. We see that the increase in the gap width shifts toward high frequencies the dip position compared to that of the peak. With regard to the phase, this same behavior can also be observed. The resonator behaves as a pure phase shifter around the resonance (see Figure 4.28(b)), and its bandwidth is equal to the frequency difference between the peak and the dip. With a simple "C" resonator, these two parameters (the peak and the dip frequency) can be controlled independently. Thus, the implementation of a hybrid coding using two independent parameters is possible.

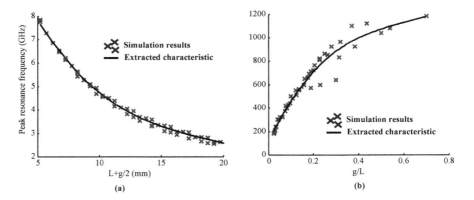

(a)

(b)

Figure 4.29. *a) The relationship between the resonance peak frequency and the length L+g/2. b) The relationship between the frequency difference between the peak and the dip and the relation g/L. The characteristics extracted in the two cases are based on 44 parametric simulations with variable L and g values*

The peak frequency is related to the path length $L+g/2$, while the separation between the peak and the dip frequencies is controlled by the relation g/L. To confirm these hypotheses, we have represented in Figure 4.29 the peak frequency as a function of $L+g/2$ and the phase deviation as a function of g/L. These values are extracted from the parametric simulations carried out under CST, by varying the length L between 5 and 20 mm and the gap g between 0.5 and 3.5 mm. On the curve in Figure 4.29(b), we observe some points that deviate from the extracted characteristic, when the relation g/L is higher than 0.2 (which corresponds to a usable bandwidth of 600 MHz).

4.6.3.2. Coding principle

The means used to code the data according to the phase and the resonance frequency of the peak are illustrated in Figure 4.30. As previously explained, by modifying the gap g, the phase rate is also modified. In addition, by modifying the length L, both peak and dip frequency are shifted. In the example shown in Figure 4.30, an ID "00" corresponds to a narrow phase deviation and a resonance peak at 2.5 GHz, while "01" is represented by a larger phase deviation by keeping the same frequency for the resonance peak. "10" and "11" codes are, respectively, related to a narrow and large phase deviation for a resonance peak at 3 GHz. In this simple example, by using a single resonator, we can code 2 bits. To increase the coding capacity, we can define more than two possible values concerning the phase deviation and the peak frequency. Thus, by using four possible values for the two resonator parameters, a capacity of 4 bits can be achieved per resonator.

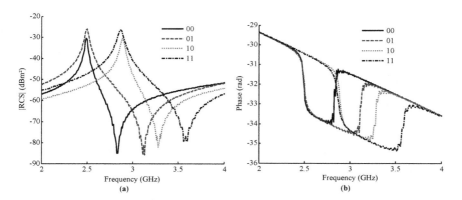

Figure 4.30. *Illustration of the hybrid coding principle that combines the use of the peak frequency position with the bandwidth for which the reflected signal is dephased by π/2: a) RCS and b) phase of the reflected signal for a distance of 100 mm between the tag and the probe. For a color version of this figure, see www.iste.co.uk/ vena/chipless.zip*

4.6.3.3. Performance and results achieved

The measurement of the eight devices has been performed in the frequency domain with the same procedure as described for the previous two designs. Tags presented in Figure 4.27 have been created on an FR-4 substrate. The allocated frequency window for each resonator is calculated as a function of the total available bandwidth and the number of resonators used. In this case, we use five resonators and we have defined a frequency band which extends between 2.5 and 7.5 GHz. Thus, each resonator has a window of 1 GHz whose initial frequency is placed, respectively, at 2.5, 3.5, 4.5, 5.5 and 6.5 GHz. The different slot and gap length values are provided in Table 4.6. The associated IDs are also provided in this table. The ID is formed by the association of P1 and P2 codes, which, respectively, represent the state linked to the phase parameter and the peak resonance frequency.

# Tag	g_1	g_2	g_3	g_4	g_5	L_1	L_2	L_3	L_4	L_5	P1 code	P2 code
1	0.5	0.5	0.5	0.5	0.5	18.4	12.7	9.7	7.8	6.4	00000	00000
2	1.5	1.5	1.5	1.5	1.5	18.9	12.7	9.4	7.4	6.1	11111	00000
3	2.5	2.5	2.5	2.5	2.5	19.1	12.5	9.2	7.1	5.7	22222	00000
4	3.5	3.5	3.5	3.5	3.5	19.1	12.1	8.9	6.7	5.4	33333	00000
5	3.5	2.5	1.5	0.5	0.5	19.1	12.5	9.4	7.8	6.4	32100	00000
6	0.5	0.5	0.5	0.5	0.5	17.7	12.7	9.7	7.8	6.4	00000	10000
7	0.5	0.5	0.5	0.5	0.5	17	12.7	9.7	7.8	6.4	00000	20000
8	0.5	0.5	0.5	0.5	0.5	18.4	11.7	9.7	7.2	6.4	00000	30400

Table 4.6. *Dimensions in mm of resonators for each associated tag and ID*

To validate this coding principle, tag 1 will be taken as a reference at a point of view of resonance frequencies and of the phase. Its ID is, therefore, "0". For tags 6–8, only the slot length is modified, with a constant gap of 0.5 mm. Regarding tags 2–5, only the gap value can vary between 0.5 and 3.5 mm by maintaining the slot lengths identical. The gap values are equal to 0.5, 1.5 2.5 and 3.5 mm, to create four different phase deviations.

Figure 4.31 presents the measurement results for tags 1 and 4 representing the extreme gap configurations, as well as for tag 5 with different gap values depending on the resonator.

First, we can note in Figure 4.31(a) that the resonance frequency of each peak is the same for different gap values g. This is possible by adjusting the length L. In fact, we have previously seen that the peak resonance frequency is linked to the term $L + g/2$. We can observe that RCS level is not the same depending on the gap and a difference ranging from 4 to 15 dB can be observed. We can also point out that the simulation results carried out for a gap $g = 0.5$ mm are very close to the measurement results, with a maximum difference of 4 dB for the peak at 2.5 GHz. Figure 4.31(b) presents the phase variations measured for these same tags. There are many significant phase variations depending on the configuration.

In Figure 4.32, the responses in amplitude (Figure 4.32(a)) and in phase (Figure 4.32(b)) of tags 1, 6 and 7 are represented. For tags 6 and 7, only the slot length of the resonator No. 1 is modified in relation to tag 1, in order to produce changes by a step of 100 MHz. Thus, the first resonance mode is shifted toward higher frequencies when the length L_1 decreases. In parallel, the phase deviation bandwidth is maintained constant with a value close to 90 MHz (see Figure 4.32(b)).

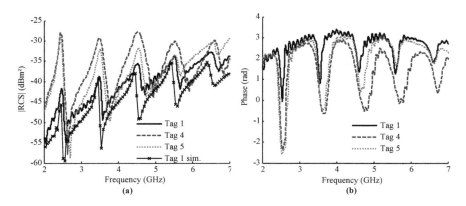

Figure 4.31. *RCS measurements for "C" tags with hybrid coding: a) in amplitude; b) in phase, for tags 1, 4 and 5. For a color version of this figure, see www.iste.co.uk/vena/chipless.zip*

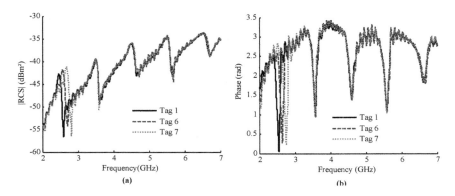

Figure 4.32. *RCS measurements for "C" tags with hybrid coding:*
a) in amplitude; b) in phase, for tags 1, 6 and 7

Measurement results which allow us to assign an ID to tags 1–8 are presented in Table 4.7. We measure for each resonance the phase deviation bandwidth. For example, for tag 4, at 2.5 GHz, we observe a phase jump of 4.5 rad. We consider the bandwidth at mid-height of this jump, i.e. at 2.25 rad and we note in this way a bandwidth equal to 181 MHz. Tag 1 contains the code "00000". This corresponds to narrower bandwidths, while tag 4 which has the largest bandwidths for all its resonators corresponds to the ID "33333". Tags 2 and 3 code intermediate IDs, as they have gap values g, respectively, equal to 1.5 and 2.5 mm. Tags 1–4 thus provide reference values for phase coding. We can compare the values measured for tags 5–8, in order to determine the P1 code for each of them. For example, the first mode of tag 5 has a bandwidth of 239.5 MHz that approaches 181 MHz of tag 4. Thus, we assign the code "3" for this mode. Modes 4 and 5 of tag 5 have a bandwidth that approaches the most the code "0" of tag 1 with a maximum error of 17.5 MHz. We can thus assign the value "0". With the same approach, we can decode the code related to the P1 parameter for the other tags.

For decoding parameter P2, resonance peak frequency values are gathered in Table 4.8. Tag 1 is the reference and its resonance frequencies must be taken as initial values. A frequency step of 100 MHz is chosen to discriminate two adjacent values. Thus, the first mode of tag 1 at 2.45 GHz codes a "0", while tags 6 and 7 with, respectively, 2.55 and 2.66 GHz code a "1" and a "2". For certain tags, we can see decoding errors. Particularly, tags 3 and 4 have undesired peak frequency shifts. In the same way, the decoding of tag 8 generates an error of assessment regarding parameter P1. To remove these errors, a more precise model taking into account the modification of the parameters of neighboring resonators is necessary.

Tag	Bandwidth measured on the phase variation for each resonator (MHz)					P1 code
	Mode 1	Mode 2	Mode 3	Mode 4	Mode 5	
1	93.5	104.5	147	133	135	00000
2	97.5	152	227	183.5	167.5	11111
3	145.5	227.5	315	284	209	22222
4	181	296.5	376.5	357	223	33333
5	239.5	249	237.5	120	117.5	32100
6	87	100	131.5	133.5	128	00000
7	92.5	89	143	127.5	140.5	00000
8	93	105	133.5	180.5	115.5	00010

Table 4.7. Measurement results for the phase parameter associated with each resonance and the corresponding P1 code

Tag	Resonance peak frequency for each resonator (GHz)					P2 code
	Mode 1	Mode 2	Mode 3	Mode 4	Mode 5	
1	2.45	3.49	4.49	5.5	6.5	00000
2	2.43	3.47	4.51	5.5	6.54	00000
3	2.41	3.48	4.55	5.53	6.57	00101
4	2.42	3.49	4.54	5.52	6.59	00001
5	2.4	3.49	4.54	5.52	6.53	00000
6	2.55	3.48	4.5	5.49	6.53	10000
7	2.66	3.47	4.5	5.49	6.51	20000
8	2.45	3.78	4.53	5.87	6.51	03040

Table 4.8. Measurement results for the frequency parameter associated with each resonance and the corresponding P2 code

Measurements show that the hybrid coding design can be implemented in practice. To estimate the number of combinations, we consider that four different phase pattern can be detected for each resonator. Concerning the resonance peak frequency, we have chosen a frequency resolution of 100 MHz. The bandwidth allocated to each resonator is 900 MHz (for example, between 2.5 and 3.4 GHz for the first mode). As a first approximation, we can calculate 10 possible frequency values in this frequency range. In reality, we must limit this frequency window to 500 MHz, to avoid covering the frequency window of the neighboring resonator, as the measured maximum phase deviation is approximately 400 MHz. Consequently, the number of states which can be coded for this second parameter is 6. In conclusion, the number of combinations for each resonator by using these two parameters is equal to $4 \times 6 = 24$. For the five resonators, the total number of combinations is equal to $24^5 = 7,962,624$, or 22.9 bits, in a surface of approximately 2×4 cm². This corresponds to a relatively large surface density coding of 2.86 bits per cm².

In Chapter 3, which is dedicated to coding, we have introduced the hybrid coding design and a graphical means to represent the coding efficiency of a symbol. In this design, the 2D constellation diagram represented in Figure 4.33 shows the improvement achieved in terms of coding efficiency. We see that on this diagram, points of physically unattainable coding appear, to avoid the frequency overlapping of neighboring resonators.

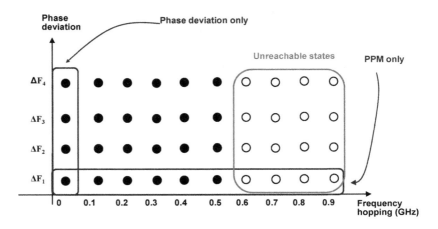

Figure 4.33. *2D constellation diagram of a "C" tag with hybrid coding*

4.6.4. *Environmental considerations, self-compensation method on the resonance frequencies extraction for tags without a ground plane*

According to the previous studies, we have seen that tags without a chip and without a ground plane can prove to be very efficient in terms of coding capacity. They have a reduced size with respect to the latest developments. This makes them compatible with low-cost applications for the identification of goods, as they can potentially be printed on the objects to identify with conductive inks. However, blocking points such as susceptibility to the environment close to the tag (whose main effect is to shift resonance frequencies) must be taken into consideration. In the following study, we implement compensation techniques [VEN 12a] of resonance frequency decodings which are related to the presence of a tag support, in order to avoid any decoding error [VEN 12d].

4.6.4.1. *Method based on the detection of the effective permittivity of the surrounding environment of the tag*

Let us use a double "C" tag (design No. 1) which was introduced earlier. The spectral signature will be considered this time, by placing the tag on an object of variable permittivity, in order to see its influence on resonance frequencies. For this, in a simulation, the tag is placed on a rectangular plate whose material has a relative permittivity ε_r and a variable thickness t. We can see its effect on the response of the tag in Figure 4.34. Figure 4.34(a) shows the influence of the relative permittivity for a constant thickness of 1 mm, while Figure 4.34(b) shows the frequency variations as a function of the thickness of the plate for a relative permittivity of 3.6. Thus, we see that all resonance peaks are shifted toward lower frequencies, particularly when the tag is placed on a cardboard box (tan $\delta = 0.1$ at 2.45 GHz).

Solutions can be implemented to manage these adverse effects. In fact, in [VEN 12d], a compensation technique based on the use of a detection resonator is presented. In the case of a double "C" tag, resonator 3 does not participate in the coding, and variations of other modes influence its resonance frequency in a less significant way. However, it is sensitive to permittivity variations of its surrounding environment, like other resonators. It is therefore possible to use its resonance frequency to trace the effective permittivity of the medium in contact, in order to deduce frequency variations of the other resonators. The frequency of this mode is 2.73 GHz when the tag is measured without a support (configuration close to the open space case). If a variation is detected on this mode which has been taken as a reference, it can be used to trace the initial resonance frequencies (i.e. without the support) of all resonators.

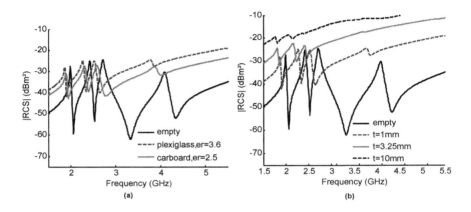

Figure 4.34. *Simulation of the frequency responses of a double "C" tag placed: a) on a support of variable permittivity (with a thickness of 1 mm), b) on a support of variable thickness for a permittivity of $\varepsilon_r = 3.6$. The substrate losses (tan δ) are 0.002 in both cases. For a color version of this figure, see www.iste.co.uk/vena/chipless.zip*

To exploit this compensation technique, it is necessary to find a relationship between the frequency deviation of the detection resonator (mode 3) and those of the other resonators. However, we hypothesize that the relation of the frequencies measured with and without a support is directly linked to the ratio of effective permittivities with and without a support, as shown in [4.7]. It is considered here that the impact of the permittivity variation is the same regardless of the resonator frequency, in any case in the considered band. In this equation, f^m_{res} and ε^m_{effres} are, respectively, the resonance frequency measured for m modes (1, 2 and 4) and their effective permittivity in the presence of a support. f^0_{res} and ε^0_{effres} are the resonance frequency and the effective permittivity obtained without a support.

$$\left[\frac{f^m_{res}}{f^0_{res}} \right] = \sqrt{\frac{\varepsilon^m_{effres}}{\varepsilon^0_{effres}}} \qquad [4.7]$$

$$\varepsilon_{effres} = k \cdot \varepsilon_{effsens} \qquad [4.8]$$

$$\left[\frac{f_{res}^m}{f_{res}^0}\right] = \left[\frac{f_{sens}^m}{f_{sens}^0}\right] \Leftrightarrow \left[\frac{f_{res}^m - f_{res}^0}{f_{res}^0}\right] = \left[\frac{f_{sens}^m - f_{sens}^0}{f_{sens}^0}\right] \qquad [4.9]$$

$$f_{res}^0 = \left[\frac{f_{res}^m \cdot f_{sens}^0}{f_{sens}^m}\right] \qquad [4.10]$$

If the tag substrate has been characterized in advance in the frequency band of interest, we can determine the effective permittivity of each resonator without the presence of a support (in this case, ε_{effres}^0). However, this is not necessary because the method described here only depends on the resonance frequency values. With regard to the effective permittivity in presence of a support marked as ε_{effres}^m, its value is unknown and varies according to the nature and the dimensions of the object on which the tag is placed. However, we can hypothesize that ε_{effres}^m and ε_{effres}^0 are linked and even proportional to the effective permittivities $\varepsilon_{effsens}^m$ and $\varepsilon_{effsens}^0$ of the detection resonator (mode 3) by a constant k, as it is indicated in [4.8]. Finally, by using [4.7] and [4.8] and by substituting the permittivities by their expression, we obtain [4.9] which establishes a direct relationship between the frequency variation of modes 1, 2 or 4 and that of mode 3 which has been taken as a reference. In this equation, f_{sens}^m is the measured frequency with the support and f_{sens}^0 is the initially measured frequency without the support (set frequency of 2.73 GHz, which is not used to code information). Finally, equation [4.10] allows us to find the initial frequency of modes 1, 2 and 4 as a function of the relative frequency deviation of mode 3. With equation [4.10], we have a means to compensate for the "detuning" effect due to an object whose nature (permittivity and losses) and dimensions are unknown. However, it must be considered that frequencies measured after "detuning" must remain detectable by the reading system which operates in the UWB band.

To verify the validity of these hypotheses [4.7] and [4.8], we have traced in Figure 4.35 the variation of the relative frequency deviation for each resonance mode in the case where a support of variable permittivity ranging from 1 to 10 is used. We note that for a permittivity of up to $\varepsilon_r = 7$, the relative frequency deviation is almost equivalent for all modes, with a maximum error of 2% between modes 1 and 3. This error will define the minimum frequency resolution that can be used. For example, at 2 GHz, a deviation of +−2% represents a frequency shift equal to +−40 MHz, while at

2.5 and 5 GHz it is, respectively, ±50 and ±100 MHz. Therefore, if the tag is placed on an object of unknown permittivity, contained in a range between 1 and 7, a frequency resolution of 100 MHz must be adopted for modes 1 and 2, while a frequency resolution of 200 MHz must be used for mode 4.

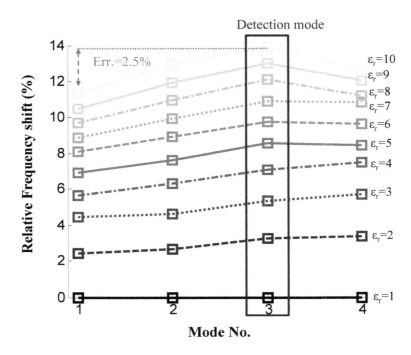

Figure 4.35. *Relative frequency shifts obtained by simulation, for each resonance mode m as a function of the support permittivity ranging from 1 to 10. Tag 1 in double "C" presented in Figure 4.18(b) has been used for this study. For a color version of the figure, see www.iste.co.uk/vena/chipless.zip*

In order to validate equation [4.10], measurements have been carried out on the size, the thickness and the variable permittivity of the supports. For this, we have used rectangular plates of PTFE, plexiglas and chipboard (Carp). With the cavity method [MEN 95], we have measured the relative permittivities, respectively, of 2.1, 3 and 4.1, as well as of tanδ of 0.002,

0.005 and 0.1 at 2.5 GHz. All rectangular plates are 1.5 mm in thickness and their size varies between 5 × 5 cm² and 10 × 10 cm². Figure 4.36 shows the recorded frequency deviations, respectively, for modes 1 and 4 before and after correction by using [4.10]. In Figure 4.36(a), the correction allows us to bring the resonance frequency value of mode 1 to the initial value with an error of less than 10 MHz for tags 1 and 4. We note for mode 4 of tag 3 (see Figure 4.36(b)), a maximum error of ±50 MHz after correction. This result remains compatible with the use of a PPM coding with a frequency resolution of 100 MHz. These results are very interesting, to the extent that significant variations of up to −280 MHz have been compensated with this approach.

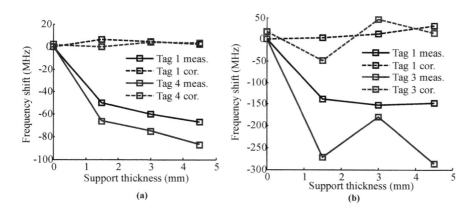

Figure 4.36. *Measured frequency shifts, related to the presence of different support thickness before and after compensation for a) mode 1 and b) mode 4. The support used is of PTFE, with a permittivity of 2.1 with losses (tant δ) of 0.002. Its dimensions are 5 × 5 cm². Tags 1, 3 and 4 (Table 4.3) have been used for this study. For a color version of this figure, see www.iste.co.uk/vena/chipless.zip*

To prove that this correction technique is reliable in practice, a measurement of a tag placed on a cardboard box filled with sheets of paper has been carried out, as we can see in Figure 4.37(a). The frequency measuring bench is the same as previously described. The calibration procedure (a procedure described in Chapter 5) requires a measurement in advance of the environment with the cardboard box without a tag, followed by the measurement of a reference tag placed on the cardboard box and whose electromagnetic response is known. The frequency response which was also measured for tag 1 is represented in Figure 4.37(b). Before correction, we can see a relatively large frequency shift for all resonance peaks, as well as a slight attenuation. After applying the correction technique on the direct measurement [4.10], resonance peaks of the corrected response are well correlated with those of the response obtained without a support. Residual frequency shifts after compensation for resonators 1–4 are, respectively, 6, 11, 34 and 0 MHz. In this case, no decoding error is recorded.

In conclusion, this correction technique can be generalized to all frequency-based single layer chipless tags. This makes their use possible even if they are placed on unknown objects. However, frequency resolution, which determines the coding capacity of the tag, is degraded due to residual errors which are still present after compensation. Thus, depending on the variation range of the permittivity of the objects, the performance varies, and as a general rule, a greater robustness in detection means a lower coding capacity.

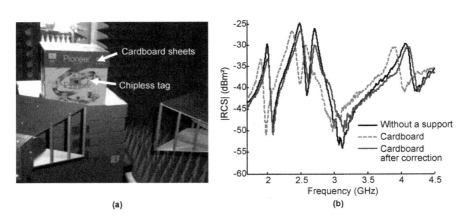

(a)   (b)

Figure 4.37. *a) Photo of a double "C" tag placed on a cardboard box containing sheets of paper; b) frequency responses obtained without and with a support before and after correction. For a color version of this figure, see www.iste.co.uk/vena/chipless.zip*

4.6.4.2. *Linear approximation method of the frequency deviation*

When we have a large number of resonators, particularly during an on-off keying (OOK) coding, a similar compensation technique can be implemented. The difference here is that it will be based on the use of two resonators acting as tags [VEN 12a]. The idea is to use the two resonators with the minimum and maximum resonance frequencies to detect the frequency variations related to the presence of a support. In Figure 4.38, we can observe the measured frequency shifts for each resonator, when tag 1 with 20 "C" resonators is placed on a plexiglas or teflon (PTFE) plate.

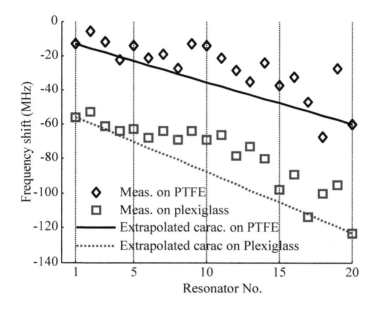

Figure 4.38. *Frequency shifts measured and extrapolated when the tag with 20 "C" resonators No. 1 (see Figure 4.23(a)) is placed on a plate of 10 × 10 cm² and a thickness of 1.5 mm², of PTFE (ε_r = 2.1, tan δ = 0.002) or plexiglas (ε_r = 3, tanδ = 0.005)*

By observing the distribution of resonance frequency points, we can note that a simple linear extrapolation based on the resonance frequencies of resonator 1 and 20 can deduce the deviations of the other resonators. A maximum error of 30 MHz is visible in relation to the extrapolated characteristic, which allows us to detect the tags without error.

Figure 4.39 presents the frequency response measured when tag 2 is placed on a cardboard box filled with sheets of paper. After correction, resonance peaks are almost confused with those of the measured response without a support. However, the attenuation effect is significant. In this case, the group delay calculated on the corrected signal (see Figure 4.39(b)) offers a remarkable detection robustness with the variation levels retained.

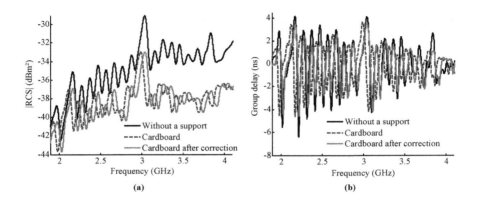

Figure 4.39. *Obtained frequency responses for the tag with 20 "C" resonators No. 2 (see Figure 4.23(b)) without and with a support before and after correction: a) in amplitude, (b) in group delay. In this case, the support is a cardboard box filled with reams of paper. For a color version of this figure, see www.iste.co.uk/vena/ chipless.zip*

Through the compensation techniques which have been presented, we see that tags without a chip and without a ground plane can be reliably detected under realistic conditions of use. For this, one or several resonators can be used to detect effective permittivity variations. This decreases the coding capacity of the tag for the same number of resonators. For example, in the case of the tag with 20 resonators that has just been mentioned, we use two resonators for "detuning" compensation, therefore we lose 2 bits of coding on the initial 20 bits. The fact remains that this technique represents an important advance, as it makes possible the development of tags printed directly on products. This will enable us to consistently reduce the manufacturing cost of chipless tags.

4.7. Design of tags with a ground plane

In the following study, we will focus on tags with a ground plane. This feature gives them different tag properties compared with those that have a ground plane, which should be exploited to improve some of their performances. We will show how it is possible to improve coding capacity by taking advantage of the fact that the created resonances have larger quality factors. Another analyzed aspect concerns the tag detection regardless of its orientation, in order to simplify the detection system. Finally, to improve the detection robustness by limiting the reflection phenomena of objects placed in the surrounding environment of the tag, a tag design based on the combination of depolarizing resonators will be introduced.

4.7.1. Presentation of design no. 4: polarization independent tag

In most of the previous designs, the polarization aspect has not been greatly considered. Consequently, to detect the tag regardless of its orientation, the reader needs to adjust the polarization of the transmitted wave to the tag orientation, which makes the architecture of the reader more complex. In the following study, we present a tag based on the combination of circular resonators [VEN 12c]. The advantage of this resonator form comes from the fact that their electromagnetic response is the same regardless of the polarization angle. The presence of a ground plane provides (1) an increased robustness of use, as the ground plane can act as a protection between the tag and the object and (2) a larger coding, since a more reduced frequency resolution can be used. Of course, this detection robustness improvement increases the unit cost of a tag, as two metal layers are necessary for its creation.

4.7.1.1. Tag description

This tag design is based on the use of a circular patch resonator. In order to optimize the occupied surface, they are embedded within each other. To ensure an identical response regardless of the polarization angle, the circular ring, which is presented in Figure 4.40, is particularly suitable. It is a known structure [CHE 82], which presents circular symmetry. Due to its form, we note that a ring of smaller radius, resonant at higher frequencies, can be embedded inside the initial ring.

When a wave arrives on the resonator, surface currents are created symmetrically along the two half-rings, defined by the wave orientation, as indicated in Figure 4.41(a). A stationary wave mode is created with a maximum value when the half-perimeter is equal to the half-length of the guided wave. The amplitude of the vertical component of the electric field is represented in Figure 4.41(b). For this tag, the substrate Roger RO4003 with a relative permittivity of 3.55 and a tanδ of 0.0027 has been used. As shown in Figure 4.41(b), the electrical fields generated inside and outside of the ring are of opposite polarity. Therefore, they interfere in a destructive way [CHE 82].

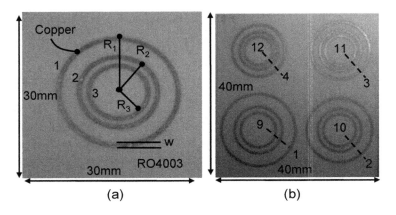

Figure 4.40. *a) Photograph of a tag composed of three embedded circular resonators. b) Photograph of a tag with 12 circular resonators. The dimensions of each circle marked as "i" are defined by its average radius Ri and its width w which is equal to 0.5 mm*

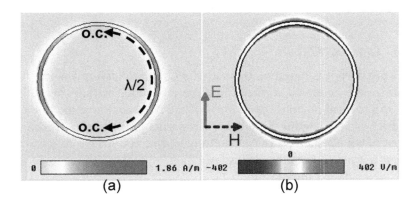

Figure 4.41. *a) Surface current density at the resonance of the design No. 4, polarization independent tag. b) Transverse electric field at the resonance. The resonator is excited by a vertically polarized plane wave. The radius of the circle is equal to 8 mm and the track width is w = 0.5 mm. The resonance frequency is 3.7 GHz. For a color version of this figure, see www.iste.co.uk/vena/chipless.zip*

To define the conductor width *w*, a compromise must be made between the space available inside the ring and the amplitude of the backscattered field. In Figure 4.42(a), we have presented the RCS of a circular patch resonator as a function of the width *w* for a substrate thickness of 0.5 mm.

Near the resonance frequency, the frequency response presents a dip, revealing a destructive interference between the structural mode and the antenna mode. A *w* value equal to 0.5 mm allows us to create a sufficiently perceptible dip in the response and leaves a relatively large surface in the interior to insert a new ring. We also note in Figure 4.42(a) that the interference dip presents a pronounced selectivity, as the rejection band at –3 dB is less than 20 MHz at 3.7 GHz. In fact, such a resonator can be considered a microstrip transmission line in the form of a half-circle, open on both sides.

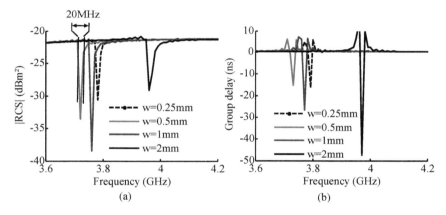

Figure 4.42. *RCS of design No. 4 (polarization independent tag): a) amplitude and b) group delay. The average radius R is 8 mm and the track width w is 0.8 mm. For a color version of this figure, see www.iste.co.uk/vena/chipless.zip*

For a low thickness substrate, the total quality factor Q [HOP 08] is mainly affected by the radiation losses, the conduction losses and the dielectric losses. Radiation losses are useful in our case, while conduction losses and dielectric losses must be as low as possible. Conduction losses are low, as copper is used, and the conductors have a thickness of 17.5 μm. However, it is necessary to choose a suitable substrate. For example, a substrate such as the FR-4 (tan δ = 0.025) cannot be used for this design. Another interesting characteristic of this resonator concerns its phase. Since the quality factor is significant, the phase change around the resonance frequency is very fast. Therefore, group delay can be significant, i.e. equal to −12 ns for w = 0.5 mm (see Figure 4.42(b)). In addition, by increasing the width w to 2 mm, this allows us to obtain a value close to −50 ns.

Equation [4.11] can be used to analytically calculate the resonance frequency of the circular patch as a function of its geometry. This expression is derived from the transmission line model [HOP 08]. The parameter R is the average radius of the ring and ε_{eff} is the effective permittivity calculated for a microstrip line with a width w, having a substrate of thickness h and relative permittivity ε_r. To take into account the presence of the electric field at the ends of the resonator, the electrical length must be slightly increased compared to the value of the half-length of the guided wave. Thus, the radius R must be multiplied by a correction factor in order to obtain a good approximation of the resonance frequency. This factor has been found with the use of a polynomial approximation based on a parametric study carried out in a simulation. Thus, [4.11] is valid for a radius R between 4 and 9 mm and a width w of 0.5 mm. Expressions similar to [4.11] can be established for other values of R and w.

$$f_r = \frac{c}{2 \cdot \pi \cdot R \cdot \left[0.965 + 19.2 \cdot R - 1372 \cdot R^2 \right] \cdot \sqrt{\varepsilon_{eff}}} \qquad [4.11]$$

The tag shown in Figure 4.40(a) is composed of three embedded rings. The frequency band has been chosen to be compatible with UWB, i.e. between 3.1 and 10.6 GHz (US regulation). Thus, the first resonator (1), i.e. the larger resonator, has a variable resonance frequency between 3.1 and 5 GHz, 5.6 and 8.1 GHz for the second one and between 8.1 and 10.6 GHz for the last one. In order to limit the coupling between the adjacent circular rings, a minimum spacing must be respected. This spacing will decrease the frequency band that can be used between two successive rings, as certain radius R values are not possible. To address this problem, at a later stage, in order to optimize the use of the available frequency band and to increase coding capacity, we can combine the resonator interleaving within each other with a side-by-side arrangement. In this way, a tag with 12 resonators such as that presented in Figure 4.40(b) can be created in a surface of 4 × 4 cm². The 12 resonators also operate in the UWB band and the available frequency range for each resonator is equal to 625 MHz.

In both cases, the coding technique used is based on a PPM coding frequency. The frequency resolution Δf which can be used in this case is 30 MHz. We have a total bandwidth of 7.5 GHz (3.1–10.6 GHz). In the first case (see Figure 4.40(a)), we have 250 frequency slots to share among three resonators. Thus, each resonator has 80 coding slots and three isolation slots (or 100 MHz to avoid any physical overlapping). Consequently, the number of combinations can be estimated at $80^3 = 512000$, or 19 bits. For the version with 12 resonators (see Figure 4.40(b)), each resonator can have 17 coding slots and three isolation slots, which provides a capacity of 17^{12} IDs or 49 bits.

4.7.1.2. Performance and results achieved

To validate this design, we have created four tags with three resonators with different configurations, as well as a tag with 12 resonators. The dimensions are provided in Table 4.9. The conductor width w is 0.5 mm and the substrate thickness is 0.5 mm. The measurement procedure of the RCS is the same as that used for the tags without a ground plane.

	R_1	R_2	R_3	R_4	R_5	R_6	R_7	R_8	R_9	R_{10}	R_{11}	R_{12}
Tag 1	8.56	5.55	4.11	/	/	/	/	/	/	/	/	/
Tag 2	8.45	5.55	4.11	/	/	/	/	/	/	/	/	/
Tag 3	8.23	5.4	4.07	/	/	/	/	/	/	/	/	/
Tag 4	7.99	5.31	4.02	/	/	/	/	/	/	/	/	/
Tag 12 res.	8.85	7.84	6.98	6.32	5.81	5.32	4.98	4.66	4.35	4.05	3.79	3.6

Table 4.9. *Dimensions of circular tags (Design n 4) in mm*

The responses in amplitude are presented in Figure 4.43. Amplitude dips between 2 and 3 dB are visible at resonance frequencies. From a point of view of the reading system, this makes the detection task difficult, and a similar amplitude noise can compromise the identification. However, as shown in Figure 4.44, group delays extracted from the phase measurement present significant variations at the resonance frequencies.

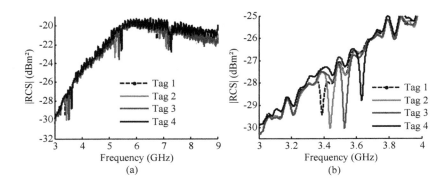

Figure 4.43. *Amplitude response of tags 1 to 4 (Table 4.6.), with three polarization independent resonators described in Figure 4.40(a): a) on the entire band and b) around the first resonance mode. For a color version of this figure, see www.iste.co.uk/vena/chipless.zip*

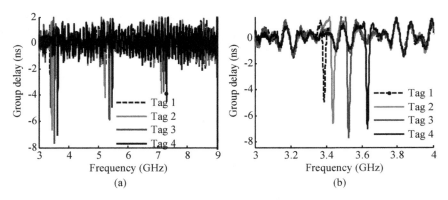

Figure 4.44. *Response in group delays of tags 1 to 4 (Table 4.6.), with three polarization independent resonators described in Figure 4.40(a): a) on the entire band and b) around the first resonance mode. For a color version of this figure, see www.iste.co.uk/vena/chipless.zip*

The obtained values are approximately −4 to −7 ns compared to −2 ns for parts of the spectrum without resonance. Thus, a detection system can simply take into account the resonance frequencies from which the group delay is lower than –2 ns. The selectivity of the peaks that appear on the representation of group delay/frequency is still better than that available in amplitude. Thus, a lower frequency resolution at 30 MHz can be adopted for the ID coding. In Table 4.10, the resonance frequencies measured for the four tags with three resonators are compared to the values obtained in a simulation and by using the analytical formula [4.11]. The obtained values are very close, which confirms that the analytical model can be used to generate the different tag configurations. A direct relationship between the resonator radius and the tag ID can, therefore, be deduced from [4.11].

	Tag 1			Tag 2			Tag 3			Tag 4		
Mode	1	2	3	1	2	3	1	2	3	1	2	3
Fr. meas. (GHz)	3.39	5.24	7.12	3.44	5.23	7.12	3.53	5.38	7.2	3.63	5.47	7.3
Fr. sim. (GHz)	3.39	5.23	7.12	3.44	5.23	7.12	3.52	5.38	7.21	3.63	5.47	7.31
Fr. calc. (GHz) Rel. [4.11]	3.4	5.23	7.12	3.44	5.23	7.13	3.53	5.38	7.2	3.63	5.48	7.29
Code	0	0	0	1	0	0	2	1	1	3	2	2

Table 4.10. *Measured resonance frequencies, simulated (CST) and calculated [4.11] for the tags with three polarization independent resonators (design No. 4)*

In terms of coding, the independence of each ring resonator is confirmed by the measurements. We can therefore assign to them an ID independently. Resonance frequencies of tag 1 are taken as references. We assign to them the code "0.0.0". Between tags 1 and 2, only the frequency of mode 1 changes from 3.39 to 3.44 GHz. We assign, therefore, the code "1.0.0" to tag 2. With the same approach and based on a frequency resolution of 100 MHz, we can find that tag 3 codes the ID "2.1.1" and tag 4 the ID "3.2.2".

To confirm the practical implementation of a tag with more than three resonators, we have measured the response of a tag with 12 resonators on a cardboard box filled with sheets of paper and the results are presented in Figure 4.45(a) for the amplitude and in Figure 4.45(b) for the group delay. It can be noted that the 12 resonances can be identified either from the amplitude or with the group delay. This last structure is particularly suitable for applications that require a relatively large coding capacity (more than 40 bits between 3 and 9 GHz).

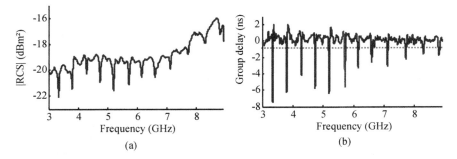

Figure 4.45. *Response measurement of the tag with 12 polarization independent resonators (design No. 4); a) in amplitude and b) in group delay*

4.7.2. *Presentation of design no. 5: polarization-coded tag*

Circular tags allow us to obtain still-unrivaled performance in terms of polarization independence and coding capacity. A design variant which was introduced in the previous section consists of opening the circular rings. Thus, we obtain split ring resonator resonators. In this case, polarization independence is no longer ensured. On the contrary, resonators become very sensitive to the incident wave polarization. This particular characteristic has been used to propose a new identification principle [VEN 12b], based on an angular coding whose main advantage is to be compatible with a very reduced bandwidth compared to frequency PPM or absence/presence coding techniques. However, the polarization sensitivity of these reflective structures allows us to create an angular sensor in a variation range of 0–180°.

4.7.2.1. *Tag description*

As a function of the incident field polarization angle, the electromagnetic response of an SRR resonator with a ground plane, as shown in Figure 4.46(a), varies significantly. From that point, it is quite possible to code an ID as a function of the tag orientation angle. For that, a more complex reader must be used by implementing a polarization agility function. However, with this technique, it is possible to obtain tags with a large coding capacity while operating only in the ISM frequency bands. This is the main driver that has motivated the study of this structure.

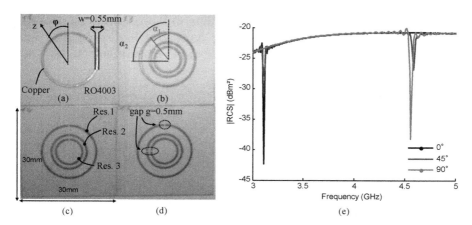

Figure 4.46. *a) Circular SRR resonator, design No. 5; b), c) and d) tags 1, 2 and 3 comprising three embedded SRRs with different configurations; e) |RCS| of a circular SRR tag as a function of frequency for several polarization angles φ. The radius R of SRR is 9.3 mm, the width w of the conductor is 1 mm and the aperture g is 0.5 mm. For a color version of this figure, see www.iste.co.uk/vena/chipless.zip*

The design presented in Figures 4.46 (b)–(d) is based on the combination of three embedded circular SRR resonators as for the previous design (design No. 4). Depending on the gap position on the ring, the response varies. The simulated electromagnetic response of a simple SRR resonator is presented in Figure 4.46(e) for several polarization angles φ (the tag remains fixed). We note that for a polarization angle of 0° (vertical polarization), the interference dip at 3.1 GHz is well visible. The resonance mode is a half-wave as in the case of a circular resonator without a gap (see Figure 4.47(a)). For a polarization angle of 90° (horizontal polarization), the resonance mode is different. In this case, the first resonance appears when the ring perimeter is equal to 3 times the half-wavelength, as indicated in Figure 4.47(b). Figure 4.46(e) shows a resonance frequency at 4.55 GHz corresponding to an electrical length of $3\lambda/2$ (the first mode in $\lambda/2$ is no longer present). For intermediate angles (45°), we can see the two resonance frequencies with variable amplitudes. However, a strong contrast is visible between an angle of 0° and an angle of 90°.

In order to generate an ID, two additional resonators are used as seen in Figures 4.46(b)–(d). The discrimination between the different resonators will occur again with frequency. Depending on the tag orientation in relation to the incident wave, and if we consider only the first resonance mode for each resonator, we can see three narrowband dips in the spectrum. However, it should be noted that the second mode can be used to increase the decoding robustness by providing a certain redundancy in the measurement.

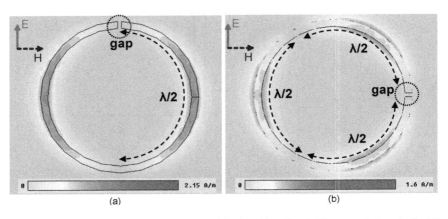

Figure 4.47. *Resonance current density (design No. 5); a) The gap is located at the top of the ring. The resonance frequency is 3.1 GHz; b) The gap is located on one side of the ring. The resonance frequency is 4.5 GHz. The resonator is excited by a vertical polarization plane wave in both cases. The average radius of the circle is 9.3 mm, the track width is 1 mm and the gap width is 0.5 mm. For a color version of this figure, see www.iste.co.uk/vena/chipless.zip*

4.7.2.2. Coding principle

The coding technique used in this approach combines a specific ID with an open ring orientation. In this way, a value of "0" is assigned to it for a rotation angle of 0° (the reference here is the location of the gap), whereas values "1","2" and "3" will be assigned to it, respectively, for 45, 90 and 135°. Thus, if we are only concerned with the amplitude variation of the first resonance mode for each ring, a minimum value indicates that the incident wave polarization angle corresponds to that of the ring (which is defined by the location of the gap). It should be noted that the rotation angle of each resonator is considered with respect to the rotation angle of resonator 1 (the largest one), taken as a reference. Thus, with resonators 2 and 3, we can code

$4 \times 4 = 16$ IDs, as it is shown in the example illustrated in Figure 4.48.

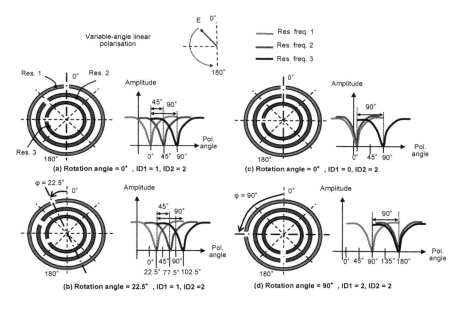

Figure 4.48. *Description of the coding technique based on the polarization diversity, design No. 5; a) The tag is not collinear with the incident wave ($\varphi = 0°$, the reference for the tag orientation is the position of the gap of resonator 1) and resonators 2 and 3 are rotated by + 45° and + 90° in relation to resonator 1; b) The tag is rotated by φ = 22.5°, resonators 2 and 3 are rotated by + 45° and +90° always in relation to resonator 1. (c) φ = 0°, only resonator 3 is rotated by +90°; d) φ = 90°, and resonators 2 and 3 are rotated by 90°. For each configuration, a curve indicates the amplitude variation of the backscattered signal as a function of the incident wave polarization and the resonance frequency of the corresponding ring. For a color version of this figure, see www.iste.co.uk/vena/chipless.zip*

ID1 is associated with resonator 2, while ID2 is associated with resonator 3. The first configuration illustrated in Figure 4.48(a) shows a tag with a rotation angle of 0°. In this tag, resonator 2 has a rotation angle of $\alpha_1 = 45°$ and resonator 3 has an angle of $\alpha_2 = 90°$. Consequently, their ID is, respectively, ID1 = 1 and ID2 = 2. Concerning the case of Figure 4.48(b), from the point of view of a reader, the tag is rotated through an angle of 22.5°. The three resonators are, therefore rotated by 22.5°. In contrast, the relative angles of resonators 2 and 3 in relation to resonator 1 remain unchanged. The ID is therefore not modified. The configurations illustrated in Figures 4.48(c) and (d) have two additional situations with another ID and a different rotation angle. It should be noted that with this coding type, even if resonance frequencies vary slightly from the presence of a disruptive environment, the ID is not modified as it depends only on the geometry of the tag regardless of the substrate used.

4.7.2.3. Performance and results achieved

To validate this coding principle, the three tags presented in Figures 4.46(b)–(d) have been created. The substrate used is ROGER RO4003 with a thickness of 0.5 mm. The radii of the circles are identical for the three tags and are, respectively, 8.56, 5.55 and 4.11 mm for resonators 1–3. The associated resonance frequencies for the first modes are, respectively, 3.4, 5.25 and 7.07 GHz. The position of the gaps varies from one tag to another. In tag 1, resonators 2 and 3 are rotated, respectively, by 45 and 90° in relation to resonator 1. In tag 2, resonators 2 and 3 are not rotated in relation to resonator 1, while in tag 3, they are rotated by 90°. RCS measurements of different tags were carried out with a frequency approach as for the previous designs. In contrast, in this particular case, the tag is placed on a rotary positioner (as it is shown in Figure 4.49), in order to obtain an electromagnetic response for each polarization angle (between 0 and 180° by a step of 5°).

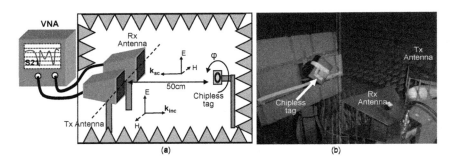

Figure 4.49. *a) Measuring bench configuration of tags – design No. 5; b) photograph of a tag under testing on the rotary positioner*

Measurement results in amplitude obtained for tag 1 and for the variable polarization angles are presented in Figure 4.50. The behavior obtained in a simulation is confirmed since we observe a strong variation in the dip amplitude as a function of the polarization. This behavior is also visible at the group delay level, as shown in Figure 4.51. In the same way as for the design introduced earlier (Design No. 4 – rings without a gap), the difference between the amplitude value obtained at resonance frequencies and that obtained for the other frequencies is significant. This greatly facilitates decoding. Figure 4.51(b) shows a magnified section around the first resonance. We obtain a well-marked amplitude dip for a rotation angle of 0° and an absence of dip for 90°.

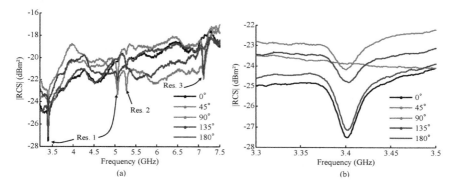

Figure 4.50. *Measured response of tag 1 (design No. 5) in amplitude: a) on the entire band and b) around the first resonance mode. For the information coding, the two configurations associated with the angles 0 and 90° are used, as they generate the greatest contrast at the reflected amplitude level. For a color version of this figure, see www.iste.co.uk/vena/chipless.zip*

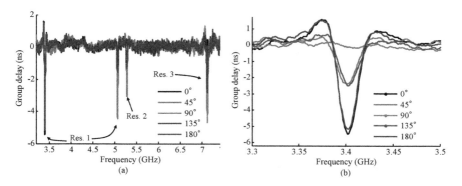

Figure 4.51. *Measured response of tag 1 (design No. 5) in group delays: a) On the entire band and b) around the first resonance mode. For a color version of this figure, see www.iste.co.uk/vena/chipless.zip*

To extract the ID of each tag, we have represented in Figure 4.52 the group delay as a function of the polarization angle of each resonator (the selection is made by choosing the corresponding resonance frequency), respectively, for tags 1, 2 and 3. These curves are extracted from the measurement results obtained for each polarization angle. For the three resonators of each tag, the first resonance modes are observed at the frequencies: Fr_1 = 3.4 GHz, Fr_2 = 5.27 GHz and Fr_3 = 7.17 GHz (see Figure 4.50(a)). The dip observed at 5.06 GHz is generated by ring 1 and presents a maximum amplitude when it is rotated by 90° as explained previously.

For tag 1, the maximum error found is 20° since we measure a rotation angle α_2 of 110° instead of 90° for resonator 3. However, by applying a filter to attenuate the rapid variations which are visible on the curve, we can easily reduce this error around 10°. For tag 2, an error of 10° is made on α_2 in relation to the theoretical values, while for tag 3 an error of 5° is made on α_1 and α_2. To measure the rotation angle of each resonator, the method comprises of searching for a minimum value. We can also search for a maximum value. In this case, it is sufficient to subtract +–90° to trace the rotations angles α_1 and α_2.

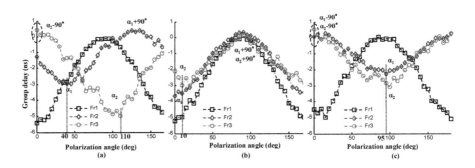

Figure 4.52. *Group times measured as a function of the polarization angle for: a) tag 1; b) tag 2 and c) tag 3. The frequencies used (measured resonance frequencies in each ring) are Fr_1 = 3.4 GHz, Fr_2 = 5.27 GHz and Fr_3 = 7.17 GHz. For a color version of this figure, see www.iste.co.uk/vena/chipless.zip*

From the obtained results, we get a maximum error of 10°. It is, therefore, possible to use a resolution of 20° for the coding of an ID.

Equation [4.12] can be used to determine the maximum number of combinations for this tag.

$$N = \left[\frac{180}{\Delta\phi} \right]^{k}$$

[4.12]

In this equation, N is the number of combinations, k is the number of resonators participating in the coding and $\Delta\varphi$ is the angular resolution. The value 180 corresponds to the maximum variation range of the polarization angle, as the information obtained following this parameter is redundant every 180°. By using [4.12], we can find a coding capacity of $9 \times 9 = 81$ combinations, or 6.34 bits.

The obtained measures confirm the high sensitivity of these polarization resonators. In addition to creating an identification function, this feature can be used as an advantage; to act as an angular sensor of any object. Due to this feature, we can even imagine detecting the angular speed of an object. At the detection system level, it is possible to create a reader operating in the ISM frequency bands. For example, it is sufficient to modify the resonator dimensions, so that they resonate in the 860–960 MHz band for the first one, 2.45 GHz for the second one and 5.8 GHz for the last one. At the measurement level, for practical reasons, we have chosen to rotate the tag around its axis of rotation. In an identification application, the tag does not rotate. Therefore, it is at the reader level that it must be able to orient the polarization of a query signal, to extract the remote tag ID. For this task, a reader implementing a polarization diversity must be developed.

4.7.3. Presentation of design no. 6: depolarizing tag

The studies presented in this section, concerning the tags with a ground plane, have been developed with the idea of improving the detection robustness in an actual environment. This point, as mentioned in the first chapter, is an essential issue to address in order to ensure the development of this technology. In fact, a chipless tag with a large amount of information, while being compact, is only of interest if it can be read in practice, i.e. in an actual environment. This aspect is not addressed in the literature. At best, we find tag measurements outside the anechoic chamber. This is a first step to show the detection robustness of a tag. However, this is not sufficient to ensure the compatibility with a practical use of a tag. In fact, we must add to that the calibration issue, which will impose a significant limit in the reading procedure, but also the question of the tag reading on different supports with the presence of other surrounding objects and the possibility of reading the tag when we know neither the object on which it is placed nor its position in relation to the reader. The study that we are going to present below is the first that provides a solution to this important issue, i.e. an actual obstacle that chipless technology is facing. We will see how the possibility of overcoming the measurement of a reference object (simplification of the measurement calibration) will allow us to resolve the problem.

For this purpose, we have developed resonators within the framework of the REP approach, which are able to depolarize the incident wave, in order to respond to an orthogonal polarization to that of a query. This provides a much better isolation between the transmitted and reflected waves and, importantly, allows us to limit the different reflections from objects surrounding the tag. In fact, in most cases, everyday objects (cardboard boxes, right support, etc.) do not have a natural tendency to depolarize the incident wave. The principle of separating the transmission and reception channel by altering the wave polarization is widely used in the field of telecommunications. It is sufficient, for example, in the case of linear polarization signals to use two orthogonally placed antennas. This principle has also been incorporated to create chipless tags containing two antennas [PRE 09b, PRE 08]. In fact, in this case, the tag is composed of a receiving antenna, a filter and a transmitting antenna that transmits a response in the orthogonal polarization at the receiving antenna. The approach introduced here is different in the sense that it is part of the framework of the REP design methodology, and it does not use antennas at the tag level. The tag design is based on multiple resonators which perform the three functions mentioned earlier. The surface required to code a large amount of information is, therefore, considerably reduced, even the presence of a ground plane is not indispensable.

4.7.3.1. *Tag description*

The polarization of the reflected wave is related to the orientation of the current paths on the conductive parts of the resonator(s). For most of the resonators, the vector sum of current densities on the metallic elements generates primarily a reflected wave with the same polarization as the incident signal. This is particularly the case for a dipole in short-circuit, a "C" resonator, a circular or a rectangular ring. To allow for the generation of a signal in a polarization other than the incident wave polarization, it is necessary to insert a dissymmetry in the geometry of the resonator in relation to the direction of the incident electric field. In Figure 4.53, we propose two structures with a ground plane which are composed of resonators that promote a cross-polarized response. The first structure uses resonators whose geometry is based on the combination of the two inverted-Ls (see Figure 4.53(a)). The resonators of the second structure in Figure 4.52(b) are composed of several dipoles in short-circuit, coupled and oriented at 45° in relation to the direction of the incident wave.

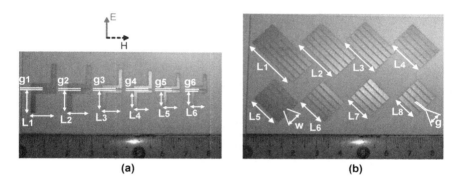

Figure 4.53. *Designs No. 6: a) depolarizing tag at double inverted-L. The dimensions are provided in Table 4.11; b) depolarizing tag at a dipole in short-circuit oriented at 45°.The dimensions are provided in Table 4.12*

When it is excited by a vertically polarized plane wave, the double inverted-L resonator in Figure 4.53(a) generates current densities which will move from the bottom to the top and from left to right or *vice versa*, as shown in Figure 4.54(a). A horizontal component (H) appears and will therefore generate a cross-polarized signal. By using two inverted-Ls rather than a single one, we can increase the RCS value while guaranteeing very good selectivity that can be controlled by the gap *g* between the two Ls. This resonator will rediffuse a part of the energy that is sensed in the vertical polarization and another part in the horizontal polarization. In Figure 4.54(b), the copolarized response (vertical polarization) presents interference dips at the resonant frequencies, while on the cross-polarized response (horizontal polarization) we observe peaks. In fact, the ground plane, like most everyday objects, does not depolarize the incident wave and this does not generate a response that would interfere with the resonant mode in the cross polarization. The structure has been optimized with the following criteria: frequency (centered in relation to the considered bandwidth), RCS level and frequency selectivity. In addition, to create the other resonators, which operate at other frequencies to achieve a PPM coding, we apply a magnification/reduction factor at an optimized resonator, which allows us to obtain a multi-resonator tag while maintaining a good selectivity even for the highest frequencies.

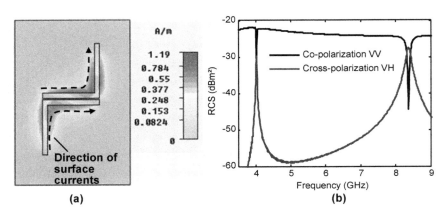

Figure 4.54. *Design No. 6, double inverted-L tag: a) resonance current density; b) frequency response simulated in copolarization and cross-polarization. The dimensions of the resonator are the following: L = 11.4 mm, g = 0.5 mm. For a color version of this figure, see www.iste.co.uk/vena/chipless.zip*

The operation of dipoles in short-circuit, coupled and oriented at 45°, is very close to that of the double inverted-L. In the same way, when a vertically polarized incident wave is sensed by the resonator, the current is guided along the dipoles. Since these dipoles are oriented at 45°, a horizontal component is created, thereby generating a cross-polarized response (see Figure 4.55(a)).

In the same way as for the double inverted-L, we observe a copolarization interference dip and a cross-polarization peak (see Figure 4.55(b)). By using multiple dipoles of the same length, we can increase the RCS level while maintaining a good selectivity. In fact, to increase the RCS, we would have to have been able to increase the dipole width, but in this case, the resonator bandwidth is increased. By using several narrow dipoles, we control both the width w and the gap g separating them, in order to optimize the resonator performance. The resonance frequency of resonators is determined by the dipole length. But unlike the double inverted-L resonator, to maintain a high RCS level for the smallest resonators, we do not modify the dipole width nor their separation. However, we lose selectivity for the highest frequencies, as we can see in Figure 4.56(b).

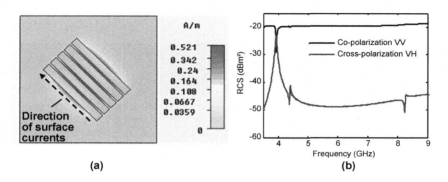

(a)   **(b)**

Figure 4.55. *Design No. 6, dipoles in short-circuit oriented at 45: a) resonance current density; b) frequency response simulated in copolarization and cross-polarization. The dimensions of the resonator are: L = 19 mm, g = 0.5 mm, w = 2 mm. For a color version of this figure, see www.iste.co.uk/vena/chipless.zip*

4.7.3.2. Performance and results achieved

For the measurement of these devices, we use the same measuring bench that is used for the tags without a ground plane, with the difference that the receiving horn antenna is rotated by 90°, in order to receive the cross-polarized response. We have created three tags for each structure on a Roger RO4003 substrate, with a thickness of 0.8 mm. The dimensions of double inverted-L tags and dipole tags in short-circuit are provided, respectively, in Tables 4.11 and 4.12.

The frequency resolution used for these designs is 100 MHz. Since we use a frequency PPM coding, we have modified certain resonance frequencies of the resonator from one tag to another, as verified in Figure 4.56.

	Tag a1						Tag a2						Tag a3					
Res.	1	2	3	4	5	6	1	2	3	4	5	6	1	2	3	4	5	6
L	11.4	10	9	8.1	7.3	6.75	11.2	10	9	8.1	7.3	6.75	10.6	10	8.6	8	7.1	6.75
g	0.5	0.45	0.41	0.37	0.33	0.31	0.5	0.45	0.41	0.37	0.33	0.31	0.48	0.45	0.39	0.36	0.32	0.31
Code	0	0	0	0	0	0	1	0	0	0	0	0	3	0	2	1	2	0

Table 4.11. *Design No. 6 – dimensions in mm of resonators for double inverted-L tags. For resonator 1, w = 2 mm*

	Tag b1								Tag b2								Tag b3							
Res.	1	2	3	4	5	6	7	8	1	2	3	4	5	6	7	8	1	2	3	4	5	6	7	8
L	25.7	21.8	19	16.8	15	13.4	12.2	11.2	24.8	21.8	19	16.8	15	13.4	12.2	11.2	23.9	21.8	19	16.8	15	13.4	12.2	11.2
Code	0	0	0	0	0	0	0	0	1	0	0	0	0	0	0	0	2	0	0	0	0	0	0	0

Table 4.12. *Design No. 6 – dimensions in mm of resonators for dipole tags in short-circuit oriented at 45° (w = 2 mm and g = 0.5 mm)*

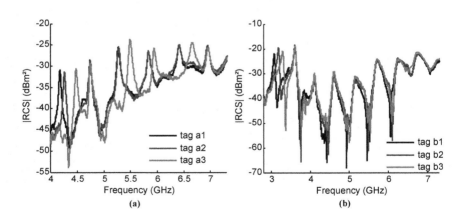

(a)   (b)

Figure 4.56. *Measured responses in cross-polarization for designs No. 6: a) double inverted-L tags (Table 4.11); b) dipoles in short-circuit oriented at 45° (Table 4.12). For a color version of this figure, see www.iste.co.uk/vena/chipless.zip*

RCS level in cross-polarization is lower for double inverted-L tags. However, we can observe that the selectivity is better, particularly toward 7 GHz. We can adopt a frequency resolution lower than 100 MHz, approximately 50 MHz for the first structure. The dipole structure at 45° allows us, however, to use a larger bandwidth, as the second resonance modes are located outside the detection spectrum. In fact, the double inverted-L structure presents a second resonant mode at 2 times the frequency of the first mode. For this reason, the initial frequency in this case is located beyond 4 GHz. For the calculation of the coding capacity of the double inverted-L, we have a bandwidth (ΔF) of 3 GHz between 4 and 7 GHz, to distribute between k = 6 resonators. Each resonator, therefore, has a frequency window Δf of 500 MHz. A resolution frequency df of 50 MHz is adopted. By using [4.13] which was introduced in Chapter 3, we obtain a number of combinations N equal to 1 million, or 20 bits.

$$N = \left[\frac{\Delta f}{df} \right]^k \qquad\qquad [4.13]$$

In the same way, we can calculate the coding capacity for the dipole structure at 45°, with eight resonators. A frequency resolution of 100 MHz and a bandwidth between 3 and 7 GHz is used. This provides a number of combinations N = 390,625, or 18.5 bits. The dipole structure is generally less efficient, as it presents a lower coding capacity at the double inverted-L structure and requires a larger surface. This is due to the worse selectivity of its resonators that requires us to choose a frequency resolution of 100 MHz.

In order to demonstrate the potential of these two depolarizing structures, we will present in the following chapter the measurement results obtained in an actual environment, i.e. outside the anechoic chamber. We will also analyze the question of the calibration procedure of the measurement.

4.8. Conclusion

In this chapter, which is dedicated to the design of chipless tags based on the REP approach, we have first introduced the general design rules. We have reviewed the different radiation properties of an electromagnetic reflector as the "C" resonator. A circuit model and an analytical model have enabled us to better understand the interference mechanism that intervenes and that is visible in the spectral response of the resonators.

This was followed by a comparative study of resonators with and without a ground plane that can be used for the creation of chipless RFID tags. These resonators are the basic elements constituting the chipless REP tags. The choice of a resonator type is crucial, as it will determine the tag performance, first in terms of surface and spectral density coding, and also from a point of view of the level of the reflected signal which is characterized by its RCS. In this classification of basic cells, "C" resonator is remarkable, because it allows us to obtain very narrow resonance peaks for a relatively small surface. However, a compromise must be made between the resonator dimensions and its RCS, so that the complete tag can be detected. The way to arrange the resonators to study their coupling has also been discussed. It is proved that a resonator frequency optimization method must be implemented to compensate for gaps that may reach 1.5% compared to the frequency of a single resonator. These differences are strongly related to the topology used to place the resonators on the substrate.

Afterward, we have presented different designs, initially discussing designs without a ground plane, then introducing tags with a ground plane. The size, coding capacity and detection robustness issues in particular have been addressed. For each design presented, several tags have been created in practice and have been measured. This has allowed us to validate the coding techniques used, as well as the levels of the reradiated signals by the tags.

In Chapter 5, we will analyze the measurement and creation aspects of the tags which are compatible with low-cost technologies. We will examine the frequency measuring bench which has enabled us to obtain the different measurements presented in this chapter. We will also introduce a temporal measuring bench based on the short pulse transmission. We will see that according to power emission masks, only this second solution seems to be viable to create a future reader. The material and creation aspects will be addressed, particularly the low-cost creation processes which are indispensable for the development of the chipless RFID technology on a larger scale.

Implementation and Measurements of Chipless RFID Tags

5.1. Introduction

In chipless Radio Frequency Identification (RFID) technology, the implementation and measurement aspects are essential elements, because they directly affect the performance of different possible designs, particularly those which were introduced in the previous chapter. We will see that by following the materials used, coding capacities and radar cross-section (RCS) values can be adjusted downward, due to the losses in the substrate or due to the resistivity of the conductive elements. Manufacturing tolerance also affects the resonance frequencies of the tags, and is a limiting parameter as far as frequency resolution is concerned. In addition, a chipless RFID tag without a robust detection system cannot have a real application. We will show how it is possible to extract an electromagnetic response of a chipless tag with a dedicated calibration procedure. First, we will discuss the implementation of a frequency measuring bench. We will also discuss the temporal characterization bench, an approach for the operation of an ultra-wideband (UWB) reading system based on radio pulse transmission.

5.2. Manufacturing process of chipless RFID tags

Chipless RFID tags can be created by simpler processes than those used for conventional RFID tags. First, there is no chip, therefore no silicon and hence a manufacturing process suitable for microelectronics. In addition, there is no connection method to connect the antenna to the chip. This contributes to the fact that a chipless tag can be more robust and less expensive than a tag with a chip. Thus, one or two maximum conductive layers are necessary, with no need for through vias to connect one layer to the other. The simplest chipless tags can even be created with the use of a single structured conductive layer placed on a substrate, like a printed circuit. However, the comparison with the printed circuits stops here. The idea here is to guide the creation of these structures toward extremely low-cost manufacturing processes. In fact, printed electronics has become a reality and so is the possibility of producing very low-cost mass tags. We will see in this chapter that the flexographic printing process which comes from the paper industry proves to be perfectly suitable for the creation of chipless tags.

5.2.1. *Manufacturing in the conventional electronics industry*

In the electronics industry, the creation of a printed circuit involves a lithographic photo-masking process, followed by chemical etching. Printed circuit manufacturers produce substrate plates with laminated and bonded copper layers on one or two sides. Next, a resin layer is created, which is photosensitive to UV light, if it is not already present on the substrates (see Figure 5.1(a)). The resin is then insolated through a mask printed on a film or on transparent glass which contains the geometric patterns that will appear on the printed circuit (see Figure 5.1(b)). The areas appearing in black will protect the resin from the UV light during insolation. However, in resin exposed to UV light, chemical bonds are altered, and it becomes soluble to developers, such as sodium. After removing the resin from the exposed areas, the substrate is immersed in an acid bath that will etch the unprotected copper. Once the copper is etched, the substrate is extracted from the bath, and then the resin that remains on the protected areas is removed with the use of acetone.

5.2.1.1. *Manufacturing limits (resolution and variability)*

This manufacturing process allows us to obtain a very good performance regardless of the designs presented in Chapter 4. However, we must be aware of the fact that the guaranteed tolerances on the dimensions of the metallic elements may prove to be problematic when frequency increases (beyond 10 GHz). In fact, the etching is called isotropic, i.e. it acts in all directions. This has a tendency to expand the gaps and to decrease the conductive track width, as we can see in Figure 5.1(b). The greater the copper thickness to be etched, the more this phenomenon will occur. The thickness of the copper, which is laminated and bonded on the substrates, can also vary by μm. In fact, substrate manufacturers provides a copper weight of m² provide as an indication, and from this information it is possible to trace the corresponding copper thickness. Etching times are, therefore, calculated in such a way that all copper can be etched. The conventional copper thicknesses on radio frequency (RF) substrates are 35, 17.5 and 9 μm. To ensure the lowest tolerance on the dimensions of the metallic elements, it is preferable to use the lowest copper thickness. In this way, etching times will be reduced and the phenomenon of track width reduction or gap increase will be limited. Currently, printed circuit manufacturers define conventionally a minimum track width of 100 μm with a tolerance of ~20 μm. This figure should be retained to the extent that it allows us to define a minimum for the frequency resolution that can be adopted for the design and coding of chipless tags. A variability on resonator lengths that is too large will have a direct impact on the initially calculated resonance frequencies.

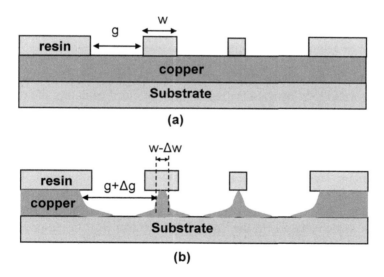

Figure 5.1. *Sectional view of an RF substrate in the etching phase: a) before etching and b) after etching. The phenomenon of gap widening and track reduction which is linked to the isotropy of the wet-chemical etching can be observed. This is even more present if we consider that the resin itself can react to the etching, or even a larger copper thickness (this is not represented here)*

5.2.1.2. *Production cost*

The production cost of a single printed circuit with a size of $20 \times 40 \text{ cm}^2$ is approximately 500 euros. The tags presented in Chapter 4 occupy an average surface of $4 \times 4 \text{ cm}^2$. It is thus possible to create approximately 50 tags for 500 euros, or 10 euros per tag. We can imagine reducing this cost by a factor of two for large quantities (or 5 euros per tag). However, we take into account the tag "configuration" aspect. In fact, if we want each tag to have a different ID, they must necessarily have a different geometry.

5.2.1.3. *Configuration*

It is difficult to imagine that for a tag topology that can code, for example, 1 million combinations, it has to create 20,000 different etching masks. It must, therefore, be considered a means to configure blank tags during a second stage after a first mass production phase. From the point of view of practical implementation, "C" tags are particularly suitable for production in two stages as described previously. In fact, from one configuration to another, a large part of the tag remains the same. To modify the resonance frequency (and therefore the tag ID), we must simply modify the length of a small short-circuit which is placed in the "C" slot.

For this, "C" resonators can potentially be mass produced. Two approaches can then be used for the configuration of each tag: the removal or the addition of material.

In the first technique, the tags must be initially created with the configuration that requires the greatest possible short-circuits (corresponding to the highest frequency for each resonator) placed in the slots. To modify the tag configuration, a mechanical milling system can be considered, to reduce the short-circuit length in a very precise way. But a faster system, such as a system based on the use of a laser, can also be considered. This technique is also mentioned several times in articles on chipless tags [PRE 09c]. Usually, the laser can remove an area of a material whose diameter corresponds to its beam, and the depth of the hole created depends on its fluence (a size expressing a power density per time unit in J/m^2). The choice of the lens at the end of the laser enables us to modify the fluence and the size of its beam. For a CO_2 laser, drilling rates of kHz [RES 16] are possible and allow us to make an ablation of the metal on a few mm and of the dielectric on a few tens of mm. Such a process allows us, therefore, to consider a mass customization of blank chipless tags. In [MOO 07], a study has been carried out on the laser drilling of a conductor and the problems related to differences in the absorption level between conductor and dielectric. A CO_2 laser is used for the drilling of copper with a thickness ranging from 3 to 105 μm. In the case of a printed circuit with a copper sheet of 17.5 μm, to avoid drilling under the conductor, the power of a laser pulse must be less than 1.2 kW for a duration of 50 μs.

The second technique consists of creating the material particularly with conductive ink inkjet printing processes. These processes will be explained in detail in the following section. In this case, the tags must be initially created with the configuration that requires the least material (initial configuration). In the case of tags with "C" resonators, this means that the slot length must be the maximum. In this case, the addition of material allows us to reduce the slot length, in order to modify the resonance frequencies of each mode.

5.2.2. *Printed electronics*

Printed electronics is a relatively new discipline, but it attracts a growing interest from manufacturers. We have seen in Chapter 3 that the basic versions of conventional RFID tags operating at 13.56 MHz can be created with the use of printed technology. However, chipless tags are much simpler to create in the sense that only a few layers of metallic ink must be deposited on a substrate.

5.2.2.1. *Conductive ink printing processes*

5.2.2.1.1. Conductive inks

There are conductive inks based on carbon particles, metal particles and conductive polymer. But currently, conductive inks based on metal particles have the best conductivity. To ensure a good conductivity, we must dry or cure the ink, in order to evacuate the solvents to strengthen the connections between metal particles. The conductive ink is composed of silver, or carbon, particles linked to each other by solving agents. After drying, these solvents disappear by evaporation. For the inks based on silver particles, a conductivity of approximately 10^5 S/m can be typically obtained (compared to 59×10^7 S/m for bulk copper). But this conductivity can be improved with an optimized annealing step. Organic inks based on graphene [HUA 11] and carbon nanotubes [DEN 09] can also be used, but the performance is lower with conductivities of approximately 10^2–10^3 S/m at best. This leads to the considerable increase in losses by conduction in the conductive material.

5.2.2.1.2. Ink jet printing

Currently, ink jet printing systems enable us to create tags in small quantities. The fundamental advantage of this technology lies in its great flexibility of use. In fact, it is not necessary to use masks to create specific geometric patterns. The operating principle of an inkjet printer is to deposit the ink through a nozzle on a precise location of the support. The print head, which is composed of a tank and the nozzle, is placed on an XY table to control the deposit position in the X and Y axes. The inkjet process that is used in printed electronics [DEA 05] is based on the DOD (Drop-On-Demand) technology. A micro ink drop comes out of the tank and goes in the nozzle without falling. The surface tension of the ink is responsible for this state. To eject the drop out of the nozzle, pressure is applied to it through a thermal or piezoelectric process. Thus, the inkjet technology is very economical in terms of ink consumption and allows an inexpensive optimization phase. However, it is not currently suitable for mass production (speed: 0.01 m^2/s), because the operating speeds are very low compared to a printing technology such as flexography. A minimum pattern size of 100 μm is a typical value in ink jet printing. Before spreading on a surface, the minimum drop size that can be ejected by the print head is typically 20 μm. Standard deposit thickness is between 1 and 5 μm. In the RFID field, research teams [KON 09] are experimenting with the manufacturing of antennas and printed circuits via ink jet printing processes with the use of specific printers [DIM 16] (see Figure 5.2(a)). The performance obtained is correct, which shows that printing technologies are suitable for the creation of chipless RFID sensors and tags. It was shown in [VEN 15b] that a chipless RFID sensor sensitive to temperature and CO_2 can be created by using multiple printing layers involving varied inks. A first ink based on silver particles can be used to create a conductive layer, followed by an ink deposit of an ink loaded with organic particles for the sensitive layer (see Figure 5.2(b)).

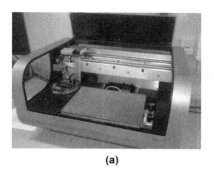

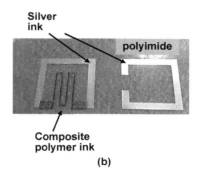

(a)   (b)

Figure 5.2. *a) Inkjet printer Dimatix Materials*
DMP-2800 optimized for electronic printing [DIM 16];
b) example of a printed chipless sensor [VEN 15b]

5.2.2.1.3. Flexographic printing

Flexographic printing is used in the printing field, as it allows us to achieve increased printing speeds, of approximately 10 m^2/s. Resolutions of approximately 50–100 μm can be achieved, which is quite compatible with the different chipless tags presented in Chapter 4. Thicker ink layers can be deposited (5 μm), which makes flexography particularly suitable for the creation of RF circuits, where very good conductivities are necessary. A flexographic printer is composed of an ink tank, in which the anilox roll (see Figure 5.3) is immersed, in order to recover ink in its micropores. The anilox roll will then transfer the ink to the cylinder on which the ink pad is bonded and which contains the printing patterns. The paper is fed between the cylinder containing the ink pad and a compression cylinder to transfer the pad pattern to paper. The paper can then be fed to a thermal curing or UV system to set the ink on its support. The amount of ink printed is adjusted with the use of an anilox roll.

For the first mass production stage, flexography is a particularly suitable technique. However, for a second "configuration" phase, the ink jet printing technique can be used due to its flexibility of use. The article [ZHE 08] on a chipless tag design coding the information in time mentions this possibility, namely the use of ink jet printing as a tag customizing method. The idea in this case is to pair the capacities at a transmission line by the addition of small localized short-circuits.

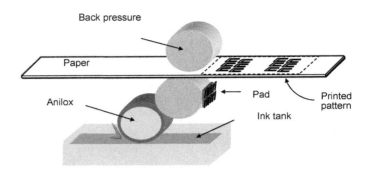

Figure 5.3. *Flexography principle*

To estimate the production cost of a printed tag, we need to know the covering surface of a chipless tag. In fact, in this kind of approach, the amount of conductive ink that should be used primarily determines the final price. Let us take, for example, a chipless "C" tag [VEN 11b] in which the effective surface of metal conductors is equal to 1 cm^2. With an anilox of 20 cm^3/m^2, we can cover a surface of 50 m^2 with one liter of ink, which represents the equivalent of 500,000 tags. With 3,000 euros/liter (in 2012), the unit cost of a tag falls to 0.6 euro cent. In the years to come, this cost will tend to decrease drastically.

5.2.2.2. Characterization of different supports

With a printing process, it is possible to use various low-cost supports, such as paper, cardboard and polymers. However, there are some inherent constraints in RF technologies that need to be considered, to ensure that printed devices can operate on ultra-high frequency (UHF) and super-high frequency (SHF) frequencies. In a dielectric, losses will limit the working frequency. They are most often expressed by the term *tan δ* which is the ratio between the imaginary $\varepsilon''(\omega)$ and the real $\varepsilon'(\omega)$ parts of the permittivity, $\varepsilon(\omega)$, as a function of angular frequency ω [5.1]. Permittivity provides information on the electrical properties of a substrate. In the expression of *tan δ* (equation [5.2]), the electrical conductivity σ is taken into account to the extent that it also participates in the losses.

$$\varepsilon(\omega) = \varepsilon'(\omega) - j \cdot (\varepsilon''(\omega) + \frac{\sigma}{\omega}) \qquad [5.1]$$

$$\tan \delta = \frac{\varepsilon''(\omega) + \dfrac{\sigma}{\omega}}{\varepsilon'(\omega)} \qquad [5.2]$$

$$Q_d = \frac{1}{\tan \delta} \qquad [5.3]$$

$$\frac{1}{Q} = \frac{1}{Q_d} + \frac{1}{Q_r} + \frac{1}{Q_c} \qquad [5.4]$$

As we have seen previously, tags without a ground plane can operate on a current substrate such as the FR-4 with a relative permittivity of 4.6 and with a *tan δ* of 0.025. However, tags with a ground plane require a low loss substrate, such as the Roger RO4003, with an approximately 10 times lower *tan δ* (0.0027). Let us now return to the quality factor notion, which was previously introduced. In fact, this factor links between the resonator frequency selectivity with the presence of losses. For frequency-coded chipless RFID tags, quality factor Q must be the greatest possible to maximize coding capacity. It is related to the dielectric losses (see [5.2] and [5.3]), the conduction losses in the conductors ($1/Q_c$) which are very low with copper and the losses by radiation ($1/Q_r$) which are preponderant in the case of resonators for chipless RFID tags. Equation [5.4] provides the expression of the total quality factor. We can note that when *tan δ* is high, the dielectric quality factor Q_d is low. The overall quality factor Q will therefore be strongly degraded by a low coefficient Q_d.

To create printed tags, it is necessary to characterize unconventional low-cost substrates. Here, we present the characteristics of several paper types, such as labels, cardboard paper and glossy paper. We also present the parameters of plastic substrates which are also compatible with the printing processes. In Table 5.1, the characteristics of different paper media obtained through the cavity method [MEN 95] are presented.

We can note that for paper substrates, *tan δ* values are approximately 0.1. Unfortunately, they are very large and do not allow us to obtain sufficient selectivities. These values are explained by the high water content of the paper and significant variations are observed in relation to its moisture content. If we keep the same structures as those described above, we should expect to adjust downward the performance in terms of tag coding capacity.

	Cardboard paper		Glossy paper		Paper (120 μm)		Latex		PE	
Parameter	ε_r	tan δ	ε_r	tan δ	ε_r	tan δ	ε_r	tan δ	ε_r	tan δ
900 MHz	2.55	0.095	3.2	0.09	3.2	0.13	2.65	0.004	2.3	10^{-3}
2.4 GHz	2.35	0.105	3	0.093	2.87	0.102	2.65	0.0027	2.55	10^{-3}
5.8 GHz	2.2	0.09	2.85	0.085	2.7	0.95	2.65	0.0027	2.55	10^{-3}

Table 5.1. *Dielectric characteristics of different printable media*

5.2.3. *Performance achieved/comparison between the different manufacturing processes*

We will now present the measurement results obtained for similar structures, carried out with a wet etching process on the FR-4 substrate and via a flexographic printing process [VEN 13b]. The media used are the common paper with a thickness of 120 μm, the cardboard paper and the glossy paper. We have, therefore, created several tags based on "C" resonators, which are described in the previous chapter. For a first print test, the dimensions of the tags have not been modified compared to the original version, i.e. those designed for the FR-4. An anilox of 20 cm^3/m^2 has been used to obtain thickness of approximately 5 μm. Ink drying at 25°C is sufficient to obtain conductivities of approximately 10^5 S/m. The structures created on different media are shown in Figure 5.4. The measurement of chipless tags was carried out with the help of the frequency-based bistatic radar characterization bench mentioned in Chapter 4. During the measurement, the tags are placed at a distance of 50 cm from the antennas.

In Figures 5.5 and 5.6, we can see the measurement results obtained, respectively, for the tags with 20 "C" elements (design No. 2) and the tags with 5 "C" resonators with hybrid coding (design No. 3) presented in Figure 5.4. To compare, we also include the measurement results for these same tags which were created on FR-4.

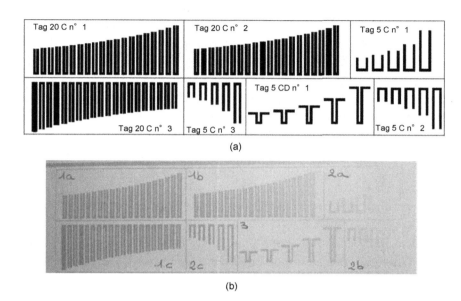

(a)

(b)

Figure 5.4. *a) Snapshot of different chipless tags created by flexographic printing; b) photograph of tags printed on cardboard paper*

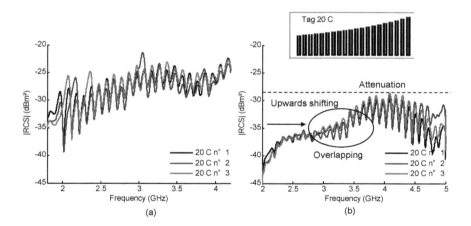

(a)   (b)

Figure 5.5. *Response measurements of three configurations of tags with 20 "C" resonators carried out: a) on an FR-4 substrate, by wet etching and b) on a 120 µm paper, by flexographic printing. For a color version of this figure, see www.iste.co.uk/vena/chipless.zip*

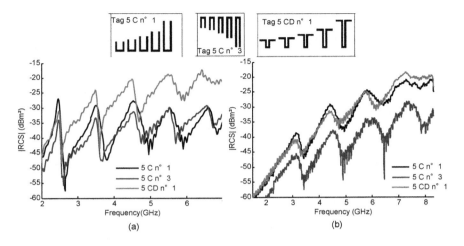

Figure 5.6. *Response measurements of tags with 5 "C" resonators carried out: a) on an FR-4 substrate, by wet etching and b) on a 120 μm paper, by flexographic printing. For a color version of this figure, see www.iste.co.uk/vena/chipless.zip*

We observe first a shift of all resonances toward the high frequencies for the paper tags. This was expected to the extent that we have kept the same dimensions as those optimized for a FR-4 substrate with a thickness of 0.8 mm. The permittivity and thickness of the paper used are lower. We observe that when resonances are visible, their level and their dynamic variation are lower. This phenomenon is particularly visible for the lowest frequencies, as we can see in Figure 5.6(b). In fact, the quality factors of lower frequency resonators are larger, therefore, the attenuation at the amplitude level is higher. Finally, we also observe a resonance peak broadening that leads, in the worst case, to the overlapping with neighboring resonance modes, as we can see in Figure 5.5(b). The tag with 20 "C" resonators is not usable as is. In contrast, for the tags with 5 "C" resonators, we observe the first four resonance modes between 3 and 8 GHz, with the last being beyond 8 GHz. The coding performance of these tags on paper is far from matching those obtained on FR-4 substrate. However, the electromagnetic signature of these tags is exploitable. A coding capacity of approximately 8 bits is possible if we consider that it is possible to code three states (corresponding to three different resonance frequencies) per resonator with the PPM frequency technique.

This first implementation phase has helped us realize the limitations induced by the use of paper substrates and the implementation processes by flexographic printing. In addition, by varying the track widths, we note that their conductivity is affected. Sometimes even cutoffs were observed on lines with a width less than 1 mm. We must, therefore, give priority to the designs with large conductive elements. However, we have found that too narrow gaps increase the chances of obtaining short-circuits unexpectedly. Moreover, a new resonator has been designed while taking these constraints into account. In order to increase the RCS, we opted for a loop resonator instead of a "C" resonator. We are referring to a loop because, unlike a "C" resonator, the metallic conductor forms a closed path (see Figure 5.7). The width of the metallic conductors has been adjusted upward, in order to increase the conductivity of the resonant elements. To ensure a good selectivity of resonators, we have placed three narrow gaps in the loop. As we have three gaps in the loop instead of a single loop, we can distribute the capacitive effect and therefore use larger spacings (i.e. less restrictive dimensions from an implementation point of view). Figure 5.7 compares the two variants. In the distributed-capacity resonator (see Figure 5.7(b)), the gap is 1.1 mm, while it is 0.48 mm for the version with 1 gap. Frequency responses are almost identical, as we can see in Figure 5.7(c). We note even an RCS level higher than 1 dB for the version with three gaps.

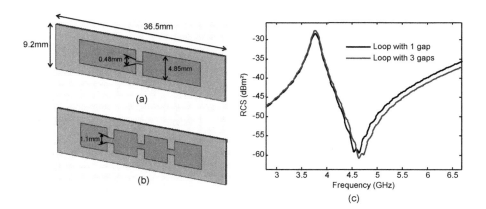

Figure 5.7. *a) Loop resonator with one gap. The parts in gray are conductive; b) loop resonator with three gaps; c) simulation results under Computer Simulation Technology (CST) of two resonators placed on a cardboard paper substrate (εr = 2.4, tan δ=0.12 at 2.4 GHz)*

We can note that this resonator occupies a surface larger than the surface for "C" resonators (4 times larger). In contrast, the performance obtained in terms of the reflection and selectivity level is significantly higher with $-27\,\mathrm{dBm}^2$ of RCS at 3.7 GHz compared to $-46\,\mathrm{dBm}^2$ for a "C" resonator with a gap of 1.5 mm. In addition, the loop resonator presents a bandwidth at -3 dB of 191 MHz compared to 240 MHz for a "C" resonator. We have, therefore, designed tags comprising of six loop resonators in the frequency band between 3 and 6 GHz (see Figure 5.8(a)).

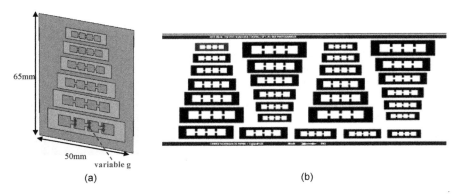

(a)   (b)

Figure 5.8. *a) Chipless tag at loop resonators for implementation on paper materials. The parts in gray are conductive; b) snapshot of flexographic printing containing four configurations of different tags and additional resonators (bottom line)*

In accordance with the RF encoding particule (REP) approach, a magnification/reduction factor has been applied on the dimensions of the resonator which has been optimized to create resonance peaks at 3, 3.5, 4, 4.5, 5 and 5.5 GHz. Four different configurations have been created (as shown in the print mask in Figure 5.8(b)) in which only gap g values vary (see Figure 5.8(a)). Tag 1 can be considered as the reference. In tags 2 and 3, the gap of resonator 1 changes, respectively, from 1.5 to 1.75 mm and 2 mm. For tag 4, all resonators are modified. The gaps are widened by 0.4 mm compared to gap values of the reference tag. Due to this, all its peaks shift toward higher frequencies.

In Figure 5.9, we present the measurement results obtained for tags printed by flexography on glossy paper with a thickness of 220 µm. In Figure 5.9(a), we note that the six resonance peaks are perfectly visible and that the response of tag 4 is clearly distinguished, as all its frequencies are shifted toward the top as expected. Figure 5.9(b) shows a larger image of the response around the first resonance frequency and again, the responses of tags 2, 3 and 4 are effectively shifted in relation to tag 1. We can note a frequency deviation of 28 MHz for tag 2, 70 MHz for tag 3 and 44 MHz for tag 4. We can, therefore, consider that frequency resolutions of approximately 50–100 MHz are quite possible, as they are detectable without ambiguity. With a frequency PPM coding and by considering a bandwidth of 3–6 GHz, a frequency resolution of 100 MHz allows us to reach a coding capacity of 13.9 bits, whereas with 50 MHz, we can achieve 19.9 bits. This design, therefore, confirms that it is quite possible to create functional tags printed on a paper substrate. For this, industrial printing techniques, such as flexography, can be employed, which allows us to achieve very large production rates.

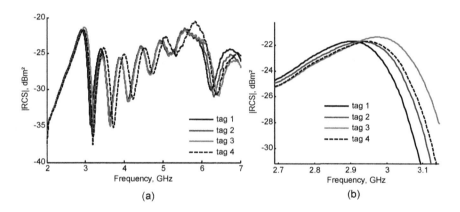

Figure 5.9. a) Measured frequency responses for four configurations of a loop tag on glossy paper (ε_r = 3, tan δ = 0.049 at 2.4 GHz, with a thickness of 220 µm); b) larger image of the area around the first resonance mode. Tag 1 has the minimum gap values. The gap of resonator 1 is reduced by 0.25 and 0.5 mm, respectively, for tags 2 and 3, and the other gap values remain the same. In tag 4, all gaps are increased by 0.4 mm in relation to tag 1. For a color version of this figure, see www.iste.co.uk/vena/chipless.zip

After having integrated the parameters of the substrate paper used for the creation of the tags in the CST simulator, we have been able to obtain simulation results that approach the measurements results, as shown in Figure 5.10.

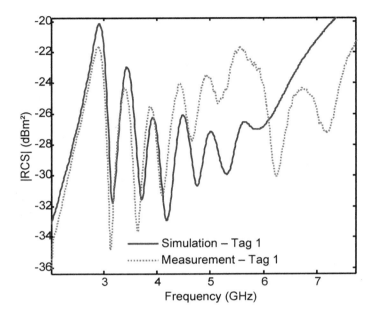

Figure 5.10. *Measurement and simulation results of a loop tag on glossy paper No. 1 (ε_r = 3, tan δ = 0.049 to 2.4 GHz, with a thickness of 220 µm). For a color version of this figure, see www.iste.co.uk/vena/chipless.zip*

It must be noted that for the higher frequencies, a gap at the peak amplitude level of approximately 5 dB is observed. This can be a result of the imprecision of the permittivity model used under CST, to model the behavior of the paper substrate at higher frequencies. Another source of probable error can be the imprecision on the measurement of the reference object and the associated analytical model.

5.3. Measurement methods of chipless RFID tags

The measurement aspect is essential in chipless RFID technology. This allows us to extract an electromagnetic signature containing the tag ID. Chipless tags are fundamentally different from conventional RFID tags. In fact, in the case of a conventional RFID, a specific frame is sent by the reader to the tag following a conventional binary amplitude modulation scheme. The tag demodulates this signal, analyzes the query, writes potentially data in its memory and returns a response by modulating its load. Chipless RFID tags operate without a communication protocol. They can be seen as radar targets with a particular temporal or frequency signature. The majority of identification applications in the conventional chipless RFID technology require a reading range of less than 1 m. This helps us ensure a certain detection robustness. The difference compared to barcodes is that we do not have to resort to a precise tag positioning opposite the reading head. Open space characterization benches, which are based on a frequency [VEN 11a] as well as on a temporal [VEN 11a, FRI 11e] approach, have been proposed. The "in contact" detection method which uses a cavity has also been introduced [DEE 10]. In fact, in some applications (for example, people or object authentication), the range is not a performance criterion. However, a reading via contact is recommended for security issues.

5.3.1. *Study of a frequency radar measuring bench*

The measurement of the electromagnetic signature of a tag, regardless of what it is, can be carried out with the use of a frequency or temporal characterization approach. In both cases, we have a bistatic configuration. The frequency approach using a vector network analyzer (VNA) is interesting for the power sensitivity and the dynamics it provides. In a frequency approach, the idea is to measure the level of the reflected response by the remote tag, as well as its phase as a function of frequency. For that, an RF source is connected to a transmitting antenna, as it is presented in Figure 5.11(a). It generates a continuous wave (CW) which performs frequency scanning. A receiver connected to a receiving antenna senses a part of the reflected signal by the radar target, as well as by the environment. At each frequency point, the ratio between the reflected and the transmitted power must be calculated, as well as their phase difference. This refers to a complex term, the S_{21} parameter, a size measured by a VNA (in this case, we assign port 1 to the source and port 2 to the receiver). Since we use two separate antennas which are separated in transmission/reception, the described assembly is thus a bistatic radar configuration [BAL 05]. By using two separate antennas for the transmission and reception, we can improve the isolation between the transmitted and reflected signals, which leads to the increase in the measurement dynamics and to the detection of weaker signals.

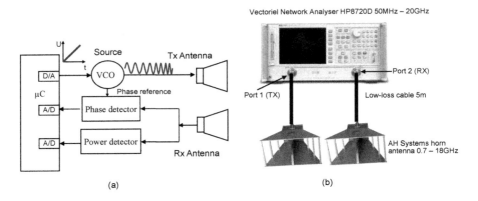

(a) (b)

Figure 5.11. *a) Synoptic view of a bistatic frequency radar bench; b) frequency measuring bench using a VNA HP8720D 50 MHz-20 GHz*

The frequency measuring bench proposed in Figure 5.11(b) is composed of a VNA allowing us to make a frequency scanning between 50 MHz and 20 GHz. Its dynamics is approximately 100 dB, which enables the sensing of minimal power ratio in relation to the transmission signal (it typically is approximately 0 dBm (1 mW)). Both antennas used are very broadband horns (0.7–18 GHz), AH Systems SAS-571, whose gain changes between 10 and 12 dBm and between 3 and 10 GHz. Measurements are carried out in an anechoic chamber, as we can see in Figure 5.12, and low-loss cables 5 m long enable us to connect the VNA to the antennas. The chipless tag to characterize is usually placed at 50 cm of each antenna on a wood or plastic support and the antennas are separated by a comparable distance, approximately 50 cm. The separation of the antennas is used to limit their direct coupling which generates a signal higher than that from the tag. To measure the copolarized response of the tag, both horn antennas are oriented in the vertical polarization, as shown in Figure 5.12(a). In contrast, to measure the cross-polarized response, the receiving horn antenna is rotated by 90°, as shown in Figure 5.12(b). In this case, we must ensure precisely that the angle is 90° in relation to the transmitting antenna, to guarantee a good isolation with the copolarized response.

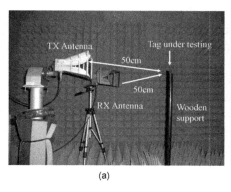
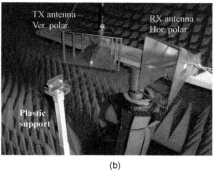

(a)   (b)

Figure 5.12. *Configuration of the anechoic chamber for a bistatic radar measurement: a) in copolarization and b) in cross-polarization*

5.3.1.1. *Calibration procedure*

Although we have not mentioned it up to this point, all frequency measurements that we have presented in this chapter and in the previous chapters have been obtained with the use of a rigorous calibration procedure. In fact, when we observe the S_{21} parameter on the network analyzer that corresponds to an empty detection environment measurement (without the tag), a level of approximately −40 to −50 dBm (for a transmitting power of 0 dBm) can be observed. This response is linked to the direct coupling between the antennas and the possible reflections of static objects in the measurement environment. By performing a power budget, we notice very quickly that this value is well beyond the power that is reflected by the tag. The latter is around −60 to −70 dBm for a distance of approximately 0.5 m. In addition, the electromagnetic signatures of the tags are configured in such a way that very slight amplitude or phase variations can code a different ID. It is, therefore, not necessary that the filtering effects of cables and antennas can compromise the integrity of these signatures. In order to better understand the measurement environment of a tag, we have represented in Figure 5.13 a transmission channel model from the source to the receiver, via the tag which acts as a filter on the signal transmitted by the source. In this model, we can link the *m(t)* measurement to the *s(t)* source by the relationship [5.5].

$$m(t) = h_2 * hf_2 * (h_{tag} + h_{sup}) * hf_1 * h_1 * s(t) + h_2 * h_{env} * h_1 * s(t) + b(t) \quad [5.5]$$

In this equation, h_1 and h_2 represent the filtering functions of cables and antennas in transmission and reception. The path traveled by the wave in both directions is represented by the transfer functions hf_1 and hf_2. The direct coupling between the transmitting and receiving antennas is represented by h_{env}. The contributions related to the presence of a tag and a support, both separated from the antennas, are represented by h_{tag} and h_{sup}. The received signal $m(t)$ therefore contains the response of these different elements that are excited by the source $s(t)$, to which a white noise $b(t)$ is added.

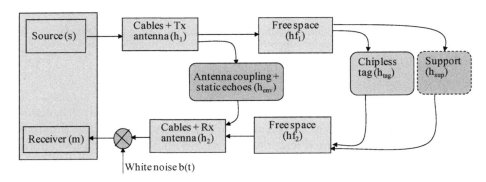

Figure 5.13. *Transmission channel modeling. The source and the receiver can be contained in the vector network analyzer. In this diagram, h tag is the transfer function to be extracted by the measurement. It varies from one tag to another. The response of the support hsup is considered constant*

Equations [5.6] and [5.7] allow us to make the link between the S_{21} parameter and the expressions of the transmission model. It should be noted that in equation [5.7] the $B(f)$ noise is random in nature. It can, therefore, be attenuated via an averaging (or median) operation, by performing several measurements. However, the acquisition time is longer.

$$S_{21} = TF\left[h_2 * hf_2 * (h_{tag} + h_{sup}) * hf_1 * h_1 + h_2 * h_{env} * h_1 \right] + B(f) \qquad [5.6]$$

$$S_{21} = H_2 \cdot Hf_2 \cdot (H_{tag} + H_{sup}) \cdot Hf_1 \cdot H_1 + H_2 \cdot H_{env} \cdot H_1 + B(f) \qquad [5.7]$$

By taking into account the communication symmetry, to simplify [5.7] we can gather the terms $H_2.H_1$ in a single term $(H_1)^2$ and the term $Hf_1.Hf_2$ by

$(Hf_1)^2$. In the end, we can obtain equation [5.8]. In this equation, we consider that the noise can be neglected by applying the method described above. $B(f)$ no longer appears.

$$S_{21} = H_1^2 \cdot Hf_1^2 \cdot (H_{tag} + H_{sup}) + H_1^2 \cdot H_{env}$$ [5.8]

To find H_{tag}, a calibration procedure must be implemented. This requires three measurements. First, the empty environment measurement is carried out without the tag and in the presence of a support, since it has been fixed. This allows us to obtain the terms of equation [5.9]

$$S_{21}empty = H_1^2 \cdot H_{env} + H_1^2 \cdot Hf_1^2 \cdot H_{sup}$$ [5.9]

$$S_{21}tag = H_1^2 \cdot H_{env} + H_1^2 \cdot Hf_1^2 \cdot (H_{sup} + H_{tag})$$ [5.10]

$$S_{21}tag - S_{21}empty = H_1^2 \cdot Hf_1^2 \cdot H_{tag}$$ [5.11]

Next, we repeat the measurement but this time in the presence of a tag and we obtain $S_{21}tag$ (see equation [5.10]). Consequently, a simple subtraction already allows us to overcome the coupling effect of the antennas, the echoes and also the support, as shown in equation [5.11]. In contrast, at this stage, we cannot yet "uncorrelate" the tag response of the filtering effect of the antennas and cables (the effects are represented by the function $(H_1)^2$), and the open space $(Hf_1)^2$. In what follows, we consider that $(H_1)^2$ is invariant, i.e. it is not modified by the presence or the absence of a tag, as well as by any object that can be placed at the measurement location of the tag. In order to obtain only the tag response H_{tag}, we must perform a third measurement [5.12], this time in the presence of a reference object whose transfer function h_{ref} is known precisely (by simulation, or analytical calculations). By making the subtraction of this measurement with that in an empty environment and by isolating the term $(H_1)^2.(Hf_1)^2$, we obtain [5.13].

$$S_{21}ref = H_1^2 \cdot H_{env} + H_1^2 \cdot Hf_1^2 \cdot (H_{sup} + H_{ref})$$ [5.12]

$$H_1^2 \cdot Hf_1^2 = \frac{S_{21}ref - S_{21}empty}{H_{ref}}$$ [5.13]

The term $(Hf_1)^2$ can be modeled in the ideal case (an open space model) by equation [5.14]. We can consider here simply a filter which attenuates and dephases the signal. If we know the distance between the tag and the antennas, we can deduce $(Hf_1)^2$. In [5.14], R is the distance between the tag and the antennas. Similarly, $(Hf_1)^2$ relating to the antennas could be measured separately. However, in the bistatic case, it depends on the configuration between the two antennas which can vary, for example if the bench is transported to be disassembled and reassembled.

$$Hf_1 = TF\left[\frac{\lambda}{4\pi R}e^{-j\frac{2\pi}{\lambda}R}\right] \tag{5.14}$$

However, it is not necessary to know $(Hf_1)^2$ if the tag is measured at the same location as the reference object.

Finally, we obtain the transfer function H_{tag} [5.15] by inserting [5.13] in [5.11].

$$H_{tag} = \frac{S_{21}tag - S_{21}empty}{S_{21}ref - S_{21}empty}H_{ref} \tag{5.15}$$

The terms in [5.15] are homogeneous in relation to the power root (exactly a power root ratio which is received on the transmitted power root). We can modify [5.15] to show the radar cross-section. In fact, RCS is defined by equation [5.16] where E_i and E_s represent, respectively, the incident and reflected electric field. We should note that these sizes are also homogeneous at the power roots. In the far-field approximation, we can therefore establish the equality [5.17].

$$\sigma = \lim_{R\to\infty}\left[4\pi R^2 \frac{|E_s|^2}{|E_i|^2}\right] \tag{5.16}$$

$$\sigma = 4\pi R^2 |S_{21}|^2 \tag{5.17}$$

Equation [5.17] can be generalized to the "complex RCS" noted as $\bar{\sigma}$ by associating the phase information. By inserting [5.17] in [5.15], we obtain [5.18] which allows us to obtain the RCS of the tag under testing $\bar{\sigma}_{tag}$ as a function of the RCS of the reference object $\bar{\sigma}_{ref}$ and three measurements carried out $S_{21} empty$, $S_{21} tag$ and $S_{21} ref$.

$$\bar{\sigma}_{tag} = \left[\frac{S_{21}tag - S_{21}empty}{S_{21}ref - S_{21}empty} \right]^2 \bar{\sigma}_{ref} \qquad [5.18]$$

This result can be found from [WIE 91] which generalizes this approach in the case of a radar system using the two copolarization terms (VV and HH), as well as the two cross-polarization terms (VH and HV). In this case, each S parameter is represented by a 2 × 2 matrix, which takes into account the potential couplings between the polarizations.

5.3.1.2. Power budget and detection range calculation

The frequency bands of RFID tags are, for amount of information purposes, very spread out and generally between 3 and 10 GHz. To create a power budget, we must be interested in this wide band side. By taking into account the procedure used, the fact that regardless of the frequency, the tags are located in the far-field, it is possible to use the radar equation to estimate the power levels that we can obtain in reception. Remember that this equation allows us to calculate the reflected power P_{rx} at the source level by a target located at a distance R. G_{tx} and G_{rx} are the receiving and transmitting antenna gains, σ represents the radar cross-section in meters and P_{tx} is the power radiated by the source. In this case, the power which is reflected and detectable by a reader is directly linked to the radar cross-section (RCS) of the tag, as it is shown in equation [5.19] [BAL 05]. An estimate of the detection range can be calculated with [5.20] which is obtained from [5.19].

$$\frac{P_{rx}}{P_{tx}} = \frac{G_{tx}G_{rx}\lambda^2}{(4\pi)^3 R^4} \sigma \qquad [5.19]$$

$$R = \sqrt[4]{\frac{P_{tx\,max}G_{tx}G_{rx}\lambda^2}{P_{rx\,min}(4\pi)^3} \sigma} \qquad [5.20]$$

For systems coding the information in the frequency domain, the information will directly be in the RCS variation as a function of frequency.

In all cases, the receiver must be able to detect a minimum RCS value. This will determine, for a given frequency and a given distance, the transmitting power and the sensitivity required in reception. If the noise level is higher than the sensitivity of the receiver or the electromagnetic response of the remote tag, we must increase the transmitting power to receive a signal above the noise level. From [5.20], we can trace a characteristic in Figure 5.14 by providing the theoretical detection range as a function of frequency and for several RCS levels of the tag between –60 and –30 dBm^2. The transmitting power is set to 0 dBm and the sensitivity threshold is set to the noise floor, measured at –80 dBm on the VNA. The antenna gain in transmission and reception G_{tx} and G_{rx} is 10 dBi.

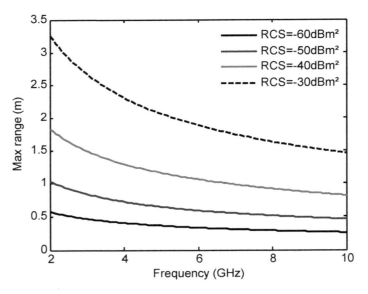

Figure 5.14. *Theoretical detection range calculated with [5.20] as a function of frequency for different RCS values of the tag. The output power (Ptxmax) is 0 dBm and the sensitivity threshold (Prxmin) is –80 dBm. The transmitting and receiving antenna gain Gtx and Grx is 10 dB. For a color version of this figure, see www.iste.co.uk/vena/chipless.zip*

In the case of tags with 5 "C" resonators (denoted as 5 C No. 3 in Figure 5.4(a)), the lowest level for a resonance peak is –45 dBm^2. In order to be able to distinguish without ambiguity, we set a dynamic of 10 dB, so that the minimum level that it must be able to detect is -55 dBm^2. Therefore, at 2 GHz, the maximum detection range is 0.7, and 0.35 m at 10 GHz. Regarding the design using the loop resonators which are described in Figure 5.8, the lowest peak level is -25 dBm^2, so we adopt a minimum RCS value to detect -35 dBm^2. In this case, detection ranges increase distinctly with 2.44 m at 2 GHz and 1.1 m at 10 GHz with 0 dBm of transmitting power. These values give us an idea of the reading ranges. However, in this case, we do not take into account the transmission mask which is defined by the Federal Communications Commission (FCC) and the Electronic Communications Committee (ECC). With a detection approach based on the transmission of a CW with a very extended frequency scanning (several GHz), the maximum authorized power spectral density is -41.3 dBm/MHz, which greatly reduces the transmitting power. We will see in what follows that the detection systems based on the transmission of ultrashort pulses can, however, allow us to achieve reading distances of approximately 1 m while respecting the regulations.

5.3.2. *Cavity measurements*

Certain applications do not require large detection ranges, but rather a tag detection at a few centimeters, or even at contact. We can think, for example, of certain people authentication applications which require a certain confidentiality level. In this specific case, confined measurement systems such as those based on the use of a resonant cavity [DEE 10] or even of a waveguide [JAN 10] can be implemented. Thus, in this book, we are interested in the characterization of chipless RFID tags in a rectangular cavity normally intended for the measurement of the low-thickness dielectric complex permittivity by the disturbance method [MEN 95] as shown in Figure 5.15. In the same way as in a mode-stirred chamber, the introduction of a metal element modifies the distributions and the frequencies of the resonance modes in a cavity. In addition, if the tag resonators have resonance frequencies close to those of the cavity, their signature will appear in the frequency response of the cavity. The resonant modes of the cavity have a very large quality factor. If we can master the resonance frequency deviation of these modes, frequency resolutions of approximately 1–2 MHz (i.e. between 20 and 50 times lower than in open space) can be achieved.

This allows us, among others, to code a larger amount of information by resonant mode in a given frequency range. With this approach, we can potentially achieve a considerably larger amount of information compared to all other investigated approaches.

To verify this design, we have created tags based on the combination of three dipoles in short-circuit and without a ground plane, as presented in Figure 5.15(c). The substrate used is the FR-4 with a thickness of 0.8 mm. With the dimensions of the cavity used, we can observe the first resonance modes, the odd TE up to 6 GHz with the use of equation [5.21]. The terms m, n, p correspond to the indications of the resonant modes according to the x, y, z axes (see Figure 5.15(a)) and the terms a, b and c are the dimensions of the cavity.

$$f_r^{TE} = \frac{1}{2\sqrt{\varepsilon_0 \mu_0}} \sqrt{\left(\frac{m}{a}\right)^2 + \left(\frac{n}{b}\right)^2 + \left(\frac{p}{c}\right)^2} \qquad [5.21]$$

The internal dimensions of the cavity that are used are $a = 203.3$ mm, $b = 38$ mm and $c = 430.4$ mm. The odd TE mode frequencies obtained are provided in Table 5.2. The measured cavity values in an empty environment are also provided for comparison.

Mode	TE101	TE103	TE105	TE107	TE109	TE1011	TE1013	TE1015	TE1017
Fr. calc. (GHz) Rel. [5.21]	0.816	1.280	1.892	2.549	3.222	3.904	4.590	5.279	5.970
Fr. meas. (GHz)	0.816	1.279	1.892	2.548	3.222	3.903	4.590	5.279	5.970

Table 5.2. *Values of TE odd first resonant modes of the Damakos cavity (see Figure 5.15(b))*

In this study, we have chosen to disrupt the odd modes TE1.0.9, TE1.0.13 and TE1.0.17 which are centered at 3.2, 4.5 and 5.9 GHz, in order to have a sufficient separation between the modes and to allow the placement of lower length resonators at the maximum dimension in the y axis (38 mm). The dimensions of the dipoles have been chosen so as to resonate around these frequencies, which provides the lengths L_1 = 23.5 mm, L_2 = 16.80 mm, L_3 = 11 mm, respectively, for resonators 1–3 (see Figure 5.15(c)). The width of the dipoles w is 3 mm and a gap g of 9 mm separates them, in order to ensure a good decoupling.

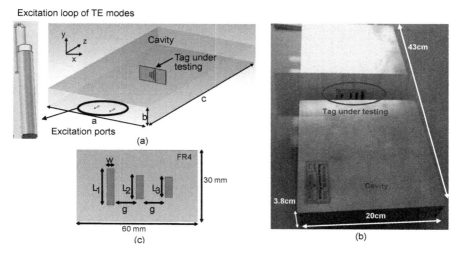

Figure 5.15. *a) Diagram of the simulated cavity under CST with the tag placed at its center; b) photo of the cavity used; c) Image of the tag with three dipoles in short-circuit of dimensions L1 = 23.5 mm, L2 = 16.8 mm, L3 = 11 mm, g = 9 mm and w = 3 mm. The thickness of the substrate is 0.8 mm*

To better understand the influence of the dipoles on the resonant modes of the cavity, we carried out a parametric study by varying the length of a dipole in simulation. Figure 5.16(a) presents the frequency responses obtained when inserting the substrate alone into the cavity or the same substrate on which a tag (resonant dipole) is placed (see Figure 5.16(b)). The substrate is used as a support to place the tag at the center of the cavity. We observe a large number of modes between 3 and 6 GHz and, among them, the modes of interest. It is difficult at first sight to identify the changes. However, if we are interested in the frequency bands around TE1.0.9, TE1.0.13 and TE1.0.17 modes, we can measure the relative resonance frequency deviation of the cavity containing a tag compared to the cavity containing a support only as a function of the length of the dipole (see Figure 5.16(b)).

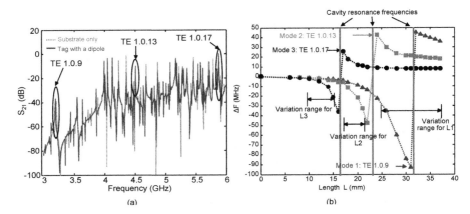

Figure 5.16. *a) Frequency response of the simulated cavity under CST with the substrate alone and the tag containing one dipole of variable length; b) relative variation (in relation to the measured frequencies with the support alone) of TE1.0.9, TE1.0.13 and TE1.0.17 modes, as a function of the length of the dipole. For a color version of this figure, see www.iste.co.uk/vena/chipless.zip*

When the resonance frequency of the dipole is close to a resonance mode of the cavity, we can observe a significant deviation on the latter. If the length of the dipole is such that its resonance frequency is greater than that of the cavity mode, we observe a shift of the latter toward the low frequencies, and vice versa. The maximum variation range for a dipole must be defined, so as to disrupt only a single resonance mode at a time. With several dipoles, we can control independently several modes and therefore increase coding capacity. From the simulation results presented in Figure 5.16(b), we can determine that dipole 1, which allows us to disrupt TE1.0.9 mode, can vary between 23 and 37 mm. While dipoles 2 and 3, which allow us to modify TE1.0.13 and TE1.0.17 modes, can vary, respectively, between 15 and 20 mm and between 8 and 14 mm.

To measure the influence of a chipless tag on the frequency response of the cavity, we proceed in the following manner. First, we conduct a measurement with the support alone placed at the center of the cavity. In addition, the resonant mode frequencies which concern us are identified. Next, we insert the chipless tag in the cavity (as we can see in Figure 5.15(b)), and the measurement is repeated. The frequency responses measured around TE1.0.9 mode for three tags with different lengths L_1 are presented in Figure 5.17. The fundamental frequency which is measured in the presence of the substrate without a dipole is 3.212 GHz. When we insert a dipole with a length $L_1 = 22$ mm, we observe that the mode is shifted by 23 MHz (resonance at 3.189 GHz). The lengths L_1 of 23 and 23.5 mm, respectively, induce a variation of 25 and 28 MHz. We therefore obtain shifts of 2 and 3 MHz, respectively, between the configurations $L_1 = 22$ mm compared to $L_1 = 23$ mm, and $L_1 = 23$ mm compared to $L_1 = 23.5$ mm. The simulation results are presented in Figure 5.17 and show a very strong similarity for the resonance frequencies between the measurements and the simulations. In contrast, in simulation, the peaks appear to be more selective, which is due to the fact that the walls of the simulated cavity have no losses, and that the actual cavity is not fully closed. These phenomena degrade rapidly its quality factor. The same behavior is observed for TE1.0.13 mode [DEE 10] with variations of approximately 3 MHz for dipole lengths which vary by 2 mm. As far as mode 3 is concerned, measurement results have not allowed us to determine the correlation between the variations of L_3 and the variation of TE1.0.17 mode. In fact, the other dipoles seem finally to affect this mode in the same way as the dipole with a length L_3.

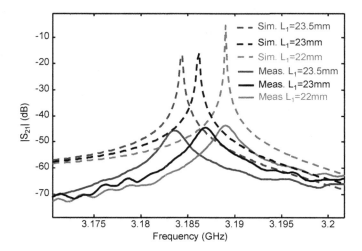

Figure 5.17. *Frequency response measurements of the cavity loaded by different chipless tags (see Figure 5.15) around TE1.0.9 mode. For a color version of this figure, see www.iste.co.uk/vena/chipless.zip*

To conclude, we can say that the cavity measurement method allows us to achieve frequency resolutions of approximately 1 MHz by the selectivity of the resonance modes of the cavity. This result is, therefore, almost 20–50 times better than what is possible to obtain with measurements of tags performed in open space. This allows us to achieve considerably higher coding efficiencies per resonator. For example, for the structure presented in this study, by using the two odd modes TE1.0.9 and TE1.0.13, and with a resolution of 2 MHz, approximately $41 \times 21 = 861$ combinations are possible providing a coding capacity of 9.7 bits. By improving the selectivity of resonators, it is possible to use a larger number of TE1.0.x modes. By using only an additional mode, generating 41 combinations, we would be able to achieve a coding capacity of $41 \times 41 \times 21 = 35.301$, or approximately 15 bits. In addition, the tag measurement at the inside of the metal cavity enables us to overcome the transmitting power problems, since the measurement is confined and allows us to achieve the confidentiality level that is required for authentication applications.

5.3.3. *Study of a temporal radar measuring bench*

We have seen previously that most frequency-coded chipless tags require a very wide bandwidth. However, it is not possible to develop a reader based on the transmission of CWs in the frequency bands between 3 and 10 GHz, because the regulatory authorities allow us to work only in the industrial, scientific and medical (ISM) radio bands. In addition, a temporal approach based on radio pulse transmission has the major advantage to obtain the entire frequency response of the tag with a single pulse. In this case, energy is concentrated on a very short duration, which allows us to obtain an instantaneous large transmitting power by keeping a low average power which is compatible with the applicable standards.

5.3.3.1. *Standards concerning the radiated powers for ultra-wideband communications*

Now, we will look into the applicable regulation. In fact, the development of chipless RFID technology is also influenced by the development of a reader, which (1) respects the standards and (2) has a price similar to the price of a standard RFID reader, i.e. around 1,000 € [VEN 15a]. The first constraint to respect lies in the power mask depending on the frequencies permitted by the regulations. This refers to the power level that it is possible to radiate in open space, without special authorization. To better understand this, Figure 5.18(a) represents the mask on the transmitting powers as defined by the FCC in the United States and the ECC in Europe. For the use of ISM bands, it is not required to obtain a license and it is possible to interrogate the tag with relatively high powers signals. However, they may not be suitable for our application. The only band that can be used to detect electromagnetic signatures spread out as those of chipless tags is the UWB. In this case, the FCC defines a transmission mask of −41 dBm/MHz between 3.1 and 10.6 GHz. The ECC is much more restrictive and imposes a very low transmission level between 4.8 and 6 GHz, as well as beyond 9 GHz. We therefore have a bandwidth of 7,500 MHz with the FCC, and of 4,700 MHz with the ECC.

If we want to comply with the UWB regulations while sending CWs with a frequency scanning, the powers that can be transmitted are extremely low, namely, −41.3 dBm maximum for a spectral spread of 1 MHz. This limits reading ranges around 10 cm. However, the values defined in the ETSI EN 302 065 standard [ETS 10] define a limit of −41.3 dBm/MHz for the calculation of an average power. For signals with very low duty cycles, as is the case for pulse radar systems, this means that relatively high instantaneous powers can be transmitted while respecting this limit in average power.

For a 7,500 MHz band, we can therefore transmit an average power of −2.54 dBm. A minimum repetition frequency (pulse repetition frequency (PRF)) is set to 1 MHz by the ECC and at 400 kHz by the FCC, in order to limit the instantaneous power. If we use a PRF of 1 MHz, or a microsecond pulse, we can calculate that with −2.54 dBm, an energy of maximum 557 pJ can be sent on 1 µs. It is, therefore, necessary to optimize the distribution of this energy in the power mask provided in Figure 5.18(a). For this, specific pulse forms such as the fifth derivative of the Gaussian pulse (see Figure 5.18(b)) can be used [ZWI 13], to respect the FCC regulation dedicated to communications inside buildings. This pulse has been defined in order to cover the 7.5 GHz of bandwidth centered around 6.85 GHz. The peak amplitude is 8 V which corresponds to an energy of 113.7 pJ, an energy lower in relation to the maximum imposed value of 557 pJ.

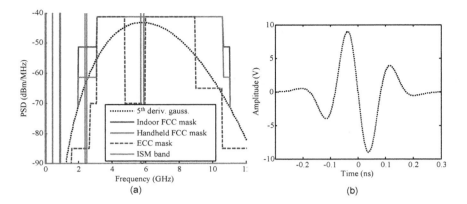

Figure 5.18. *a) Transmission mask defined by the ECC and the FCC and power spectral density of the 5th derivative of a Gaussian pulse; b) temporal characteristic of the 5th derivative of a Gaussian pulse. For a color version of this figure, see www.iste.co.uk/vena/chipless.zip*

The signal that allows us to cover this bandwidth is spread out universally on a time T_p of 600 ps on a period of 1 μs. The duty cycle is, therefore, $600.10^{-12}/1.10^{-6}$ or 1 for 1667. The gain on the report G_{RC} [5.22] which is relative to this short interval is, therefore, 1667. This corresponds to +32.2 dB in relation to the power calculated on a time T_p of 1 μs. Thus, [5.23] allows us to calculate the power of a transmitted signal P_{tx}^{TP} in a short period T_p from the average power P_{tx}^{moy} of this signal and the gain G_{RC}.

$$G_{RC} = \frac{1}{PRF \cdot T_p} \qquad [5.22]$$

$$P_{tx}^{TP} = P_{tx}^{moy} G_{RC} \qquad [5.23]$$

However, we must consider that, contrary to the impulse-radio UWB communication systems, the duration of the tag response is not equal to the duration of the source pulse. In fact, captured energy is restored on a period greater than the duration of the transmission pulse. To estimate its duration, we can use the expression of the quality factor [5.24]. Temporal response is much longer when the resonance frequency is low. In addition, we consider the minimum frequency of the UWB, namely, 3.1 GHz. A resonant mode with a bandwidth of 50 MHz (which is typically used in chipless RFID technology) presents according to [5.24] at a frequency of 3.1 GHz, a quality factor of 62. As we have seen in Chapter 4, resonators can be modeled by the second-order transfer function. For a second-order resonant system, we can match a response time $Trep_{n\%}$ at n % of the final amplitude, as a function of the quality factor Q by using equation [5.25]. For a response time at 5%, a resonance frequency of 3.1 GHz and a quality factor of 62, we obtain a value equal to 19 ns. We can conclude that the majority of resonant modes which are created by the chipless tags operating in UWB restore most of their energy in less than 20 ns.

$$Q = \frac{\Delta f}{f_0} \tag{5.24}$$

$$Trep_{n\%} = \frac{Q}{\pi f_0} \log_e (100 / n) \tag{5.25}$$

In order to propose a power budget based on the use of the radar equation, we must use average powers defined on the same time window. This time window with a duration T_p is here equal to 20 ns, or the necessary duration for the restoration of almost all of the energy of the pulse from the source by the resonators of a chipless tag. By using [5.22] with a time T_p of 20 ns, we obtain a power gain of 50 or 17 dB. A power budget can be calculated by using the formulation of the radar equation [5.20], in order to incorporate the average transmitted power in the entire UWB band denoted as $P_{tx\ eirp\ UWB}^{moy}$ and the duty cycle gain G_{RC} [5.26]. According to the standard, the maximum value of $P_{tx\ eirp\ UWB}^{moy}$ equals to -2.54 dBm. For the pulse described in Figure 5.18 (5th derivative of a Gaussian pulse), $P_{tx\ eirp\ UWB}^{moy}$ equals -9.44 dBm. Finally, by inserting [5.23] in [5.26], we obtain the expression of the theoretical detection range which depends on the pulse repetition frequency PRF and the observation window of the reception signal T_p [5.27]. In this equation, we directly express the Equivalent Isotropic Radiated Power (EIRP) power, thus the transmitting antenna gain no longer appears. However, the receiving antenna gain is always present and represents an adjustment variable of the system, to increase the detection range in the case of a bistatic radar configuration.

$$R = \sqrt[4]{\frac{P_{tx\ eirp\ UWB}^{moy} G_{RC} G_{rx} \lambda^2}{P_{rx\ min} (4\pi)^3} \sigma} \tag{5.26}$$

$$R = \sqrt[4]{\frac{P_{tx\,eirp\,UWB}{}^{moy}G_{rx}\lambda^2}{PRF \cdot T_p \cdot P_{rx\,min}(4\pi)^3}\sigma}$$ [5.27]

With the previously defined pulse, i.e. the one that is compatible with the FCC regulation, we have traced the theoretical ranges in Figure 5.19 which can be achieved by applying [5.27]. The receiving antenna gain is set to 10 dB.

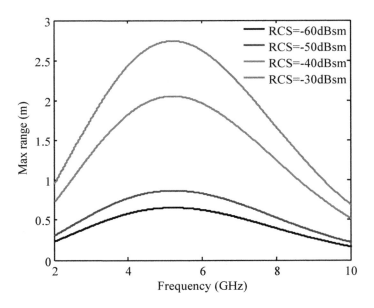

Figure 5.19. *Theoretical detection range (expression [5.27]) as a function of frequency for RCS values set to −60, −50, −40 and −30 dBm². The pulse of the Figure 5.18 is transmitted toward the tag. The noise floor Prx min is set to −80 dBm. For a color version of this figure, see www.iste.co.uk/vena/chipless.zip*

We can observe that the ranges between 1 and 2 m can be achieved on the entire frequency band for an object with a minimum RCS of −40 dBm². It is a very interesting result, as it demonstrates that a temporal radar system whose transmitting power between the regulation imposed by the FCC allows us to achieve detection ranges higher than 1 m [VEN 15a]. In addition, the design of a reader based on this operating principle is quite possible, as we will see later in this chapter.

5.3.3.2. *Characterization bench with the use of pulse signals*

The bistatic radar configuration is retained. In addition to ensuring a good isolation between the signals transmitted and backscattered by the tag, this architecture has the advantage of being able to insert a low-noise amplifier in the reception chain. In fact, an amplifier becomes essential in this case if we want to increase the sensitivity of the sampler and therefore the reading distance of the system. A highly directional antenna can also be used in reception.

The antennas are identical to those used for the frequency measuring bench, as we can see in Figure 5.20(a). The pulse generator Picosecond Lab 3500D generates a Gaussian pulse with a width of 70 ps at mid-height. It delivers a maximum amplitude of 8 V under 50 ohms. This provides a maximum instantaneous peak power of 31 dBm. This pulse type has a wide band frequency signature and allows us to cover the frequency range of 0–7 GHz. A record of the pulse form is presented in Figure 5.21(a) and its associated power spectral density in Figure 5.21(b). In reception, the oscilloscope used is the Agilent DSO91204A (see Figure 5.20(b)) whose analog bandwidth is equal to 12 GHz and the sampling rate in real time is 40 Gs/s. Its analog-to-digital converter has a resolution of 8 bits for a zero frequency signal. However, by increasing the working frequency at a few GHz, the effective resolution decreases to 4.5 bits [TEK 09], because the least significant bits are drowned in the noise.

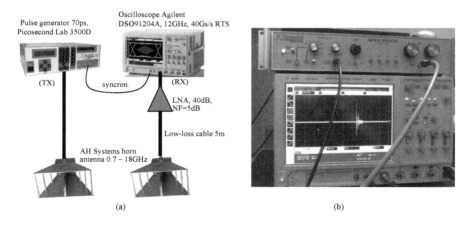

Figure 5.20. *a) Temporal measuring bench, 0–7 GHz; b) image of the pulse generator and the oscilloscope displaying the tag response*

To artificially increase the number of effective bits and to optimize the sensitivity, an averaging based on several measurements can be carried out. Thus, with 64 records, the number of effective bits of the analog-to-digital converter changes from 4.5 to 6 bits for a full scale of 40 mV. This provides a power sensitivity of approximately –50 dBm. By adding an amplifier of 40 dB in the reception chain, we increase this sensitivity to around –90 dBm. However, the output power of the low-noise amplifier is 15 dBm, and we have observed that when the antennas are too close and the pulse amplitude is set to 8 V, the amplifier is saturated at the output by an extremely strong signal resulting from the coupling between the two antennas. In this case, we must either increase the distance between the antennas (in other words, to improve the decoupling between the two antennas) or attenuate the signal between the receiving antenna and the amplifier. In fact, contrary to a vector network analyzer, an oscilloscope has a very reduced dynamic range, which is principally limited by the resolution of its sampler, 8 bits, which provides approximately 40 dB. It is therefore difficult to distinguish at the same time a high amplitude signal and a low amplitude signal.

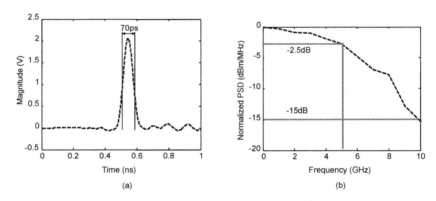

Figure 5.21. *a) Pulse measurement at the output of the Picosecond generator; b) standardized power spectral density of the same pulse*

Raw signals which are sampled by the oscilloscope at the output of the low-noise amplifier and from an empty chamber measurement of a chipless tag and a metal plate are presented in Figure 5.22(a). A larger image of the same measurements carried out around the first echo is provided in Figure 5.22(b). We can note that a large part of the signal is due to the direct coupling between the two antennas. A few nanoseconds after the coupling signal, we observe the response of the plate. This moment can be used later as the initial time for the impulse response of chipless tags. In fact, a time gating will be applied from this moment.

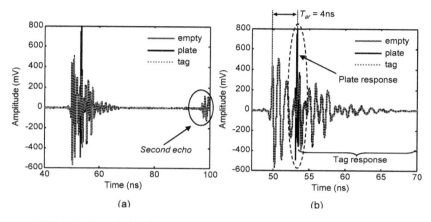

Figure 5.22. *a) Copolarization measurement of the raw temporal response of the empty chamber, of a metal plate of 5 × 5cm² and of a chipless tag of 2 × 4 cm² without a ground plane; b) Larger image of the impulse response around the first echo. For a color version of this figure, see www.iste.co.uk/vena/chipless.zip*

However, to eliminate the different echoes of static objects in the surrounding area, as well as from the support that acts in the same time window as the tag and which will therefore be superimposed, a calibration procedure is still necessary. Without this, it is difficult to extract the tag response. For example, we see in Figure 5.22(b) that the tag response appears to be perfectly superimposed in an empty environment (or even at the plate response outside of a very short interval at 4 ns after the beginning of the signal). However, when we subtract the difference between the two, we see that the tag response is present, but is very low compared to the other contributions.

The previously introduced calibration in the case of a frequency measurement remains valid and the transmission channel modeling in Figure 5.13 can still be used. However, equation [5.18] should be rewritten to show the measured voltages and not the S parameters. In fact, a frequency power ratio corresponds to a deconvolution product of the signal measured in reception $m(t)$ (expressed in Volts) in relation to the source $s(t)$. By applying the Fourier transform to this deconvolution product, we obtain the S_{21}

parameter. Equation [5.28] incorporates these modifications with $m_{tag}(t)$, the recorded temporal signal in the presence of a tag, $m_{empty}(t)$, the signal of the empty chamber, and $m_{ref}(t)$, the recorded signal in the presence of the reference object. Finally, we can deduce equation [5.29]. The terms $TF[s(t)]$ are removed.

$$\bar{\sigma}_{tag} = \left[\frac{TF[m_{tag}(t) *^{-1} s(t)] - TF[m_{empty}(t) *^{-1} s(t)]}{TF[m_{ref}(t) *^{-1} s(t)] - TF[m_{empty}(t) *^{-1} s(t)]} \right]^2 \bar{\sigma}_{ref} \qquad [5.28]$$

$$\bar{\sigma}_{tag} = \left[\frac{TF[m_{tag}(t)] - TF[m_{empty}(t)]}{TF[m_{ref}(t)] - TF[m_{empty}(t)]} \right]^2 \bar{\sigma}_{ref} \qquad [5.29]$$

Figure 5.23(a) shows the result of a simple subtraction of the $m_{tag}(t)$ signal registered for a double "C" tag with the signal from the empty chamber $m_{empty}(t)$. The spectral response associated with this temporal signal is presented in Figure 5.23(b). We will see that the first three resonance peaks appear. However, the fourth peak is less clear. In fact, the overall signal rate is distorted by the cables, the antennas and the power spectral density of the transmission signal, which is not constant between 1 and 6 GHz. By applying the calibration procedure described by equation [5.29], we obtain the impulse response of Figure 5.24(a) and the associated spectral response in Figure 5.24(b). We see that the obtained response is very close to the simulation results.

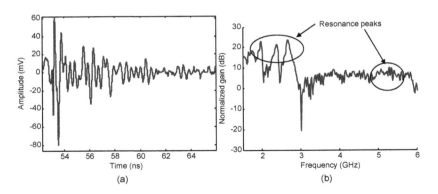

Figure 5.23. *a) Uncalibrated impulse response measurement of a double "C" tag operating between 2 and 5.5 GHz; b) uncalibrated corresponding frequency response. The first three resonance peaks are well visible*

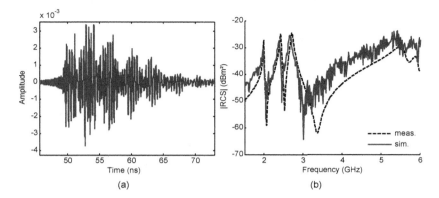

Figure 5.24. *a) Calibrated impulse response measurement of a double "C" tag operating between 2 and 5.5 GHz; b) calibrated corresponding frequency response. All resonances are well visible*

With the use of this calibration procedure, we have measured different chipless tags. We present in Figure 5.25 the response for double "C" tags [VEN 11a], as well as depolarizing tags.

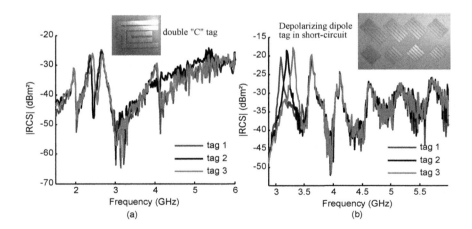

Figure 5.25. *Measurements of the spectral response obtained from temporal measurements: a) for double "C" tags; b) for depolarizing tags based on a dipole in short-circuit. Tags are arranged at 1 m from the antennas. For a color version of this figure, see www.iste.co.uk/vena/chipless.zip*

Other devices, such as metal letters, have also been able to be measured with this temporal radar bench [VEN 11c]. We observe that in most cases, when the resonant peaks are selective in frequency, the measured frequency response is very close to that obtained in simulation. The results of these measurements demonstrate that the temporal approach is quite relevant. It is in fact the key to the problem, to the extent that it is the only approach that allows us to obtain detection ranges of approximately 1 m, while respecting the UWB transmission mask imposed by the FCC or the ECC.

5.3.4. *Design of a reader of chipless tags*

The temporal radar bench has shown its effectiveness regarding the electromagnetic signature detection of chipless tags. However, in order to enable the deployment of chipless RFID technology, it is essential to propose a portable or transportable, and especially low-cost, detection system. The temporal measuring bench architecture consisting of a bistatic radar with a real-time sampler is presented in Figure 5.26(a). Previously, we have used a generator transmitting a pulse of 70 ps with a repetition frequency of 1 MHz, two horn antennas, an analog bandwidth oscilloscope at 12 GHz with a sampling rate in real time at 40 Gs/s. We have also increased the sensitivity in reception with a low-noise amplifier. Taken separately, as laboratory equipment, they are very expensive (approximately 100,000 € in total). In addition, there is not really an inexpensive alternative that allows us to achieve a sampling rate of 40 Gs/s. At best, we can find analog-to-digital converters at 2 Ge/S that should be parallelized through a complex architecture and a very fast logical device which is capable of handling huge data streams.

5.3.4.1. *Equivalent-time sampling*

We observe that unlike a conventional communication diagram in which the data passing through the RF link are not repeated, a chipless tag is a completely passive radar target which, when it is stationary, will always provide the same RF signature as a response. An architecture based on an equivalent-time sampling (see Figure 5.26(b)) is therefore possible.

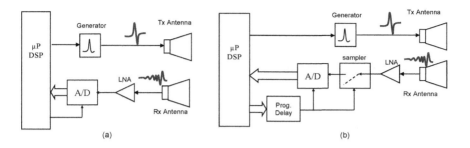

Figure 5.26. *Synoptic view of a bistatic temporal radar: a) with real-time sampling and b) with equivalent-time sampling*

The interest of this second approach is to no longer require the use of fast analog-to-digital converters. Only a sampler with an analog bandwidth of the same order of magnitude as that required for the chipless tag detection is necessary. This implies that the sampling frequency can be very low (a few MHz). However, the time necessary to block a reception signal sample must be approximately a few picoseconds. Finally, this approach allows us to sense a signal with a bandwidth of a few GHz with a sampling frequency of a few MHz. The basic principle of equivalent time is to reconstruct the signal received from several measurements whose sampling time is delayed from one measurement to another. This delay will correspond to the equivalent temporal resolution that can be obtained. In fact, the obtained resolution is a function of the increment of time between each measurement and does not correspond, as in a real-time acquisition, to the inverse of the sampling frequency. Figure 5.27 illustrates the principle of equivalent time. In this example, the signal reconstruction is performed in four stages (i.e. four pulse sending actions and four acquisitions) by incrementing a delay dt at the sampler clock level between each pass. To achieve the oscilloscope performance described above, there must be a programmable delay line with a resolution of 25 ps. With a sampling frequency of 1 MHz, we acquire a sample every microsecond. Therefore, to cover an RF signature with a duration of 20 ns, with a point every 25 ps in equivalent time, we should have 800 points, or a duration of 0.8 ms. The necessary time increase to acquire a signal is the main disadvantage of equivalent time. In addition, in order to increase the signal-to-noise ratio, an averaging operation on 64 measurements is typically performed with the temporal measuring bench.

Finally, chipless tag detection can be performed in 64 × 0.8 = 51.2 ms with an equivalent-time method, which is still relatively fast. A possible way to detect the temporal response of a tag is to divert the use of a commercially available UWB pulse radar, which is normally achieved through the walls of localization operations [RAM 16].

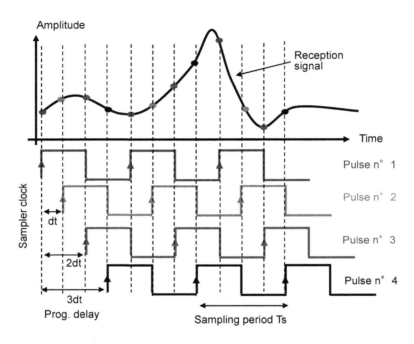

Figure 5.27. *Equivalent-time sampling principle. The reception signal is obtained in four stages. Each acquisition removes three different points of the curve. The signal transmission must be synchronized with the sampler clock. For a color version of this figure, see www.iste.co.uk/vena/chipless.zip*

5.3.4.2. *UWB pulse radar modified for chipless tag detection*

UWB radar systems for the detection of objects or persons through the walls are commercially available. These systems have the advantage of being much more inexpensive (< 10,000 €) compared to the temporal measuring bench which was presented previously. However, they are not really dedicated to the reading of chipless tags. What we seek in a location system is primarily the accurate determination of time-of-flight of the signal in both directions and not the evolution of the waveform reflected by an object. However, certain products offer the possibility to display the waveform of the signals sensed by the receiving antenna [RAM 16]. It is the essential prerequisite to transform a simple localization radar to a chipless RFID tag reader. The radar proposed by Novelda [NOV 16] is composed of an RF front end created in Complementary Metal Oxide Semiconductor (CMOS) technology and placed on an electronic board, as is shown in Figure 5.28. In this RF front end, a pulse generator and a sampler operating on a principle of equivalent time, different from the one previously explained, are embedded. In fact, the latter is asynchronous [VU 10] and the return channel has a 1-bit analog-to-digital converter battery. On the electronic board, a microcontroller enables us to control the RF front end and to make a connection with a PC on which a localization application can operate. We replaced this application by a data acquisition program under Matlab that allows us to extract and process the raw signals which are recorded by the RF front end. Next, we replaced the basic antennas which are delivered with the radar to connect the transmission/reception part to the horn antennas that were used during the previous measurements, as shown in Figure 5.28. This radar is designed to operate in a frequency band ranging from 0.45 to 9.55 GHz and its equivalent-time sampling frequency is greater than 30 Gs/s.

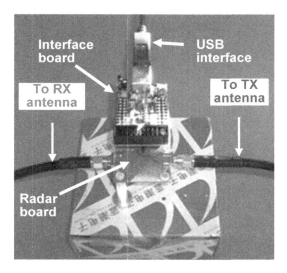

Figure 5.28. *Image of the Novelda radar connected to the horn antennas which are placed in the anechoic chamber*

To verify these characteristics, we have recorded the transmission pulse (1) on the wideband oscilloscope (see Figure 5.29(a)) and (2) on the radar with a loopback from the transmitter to the receiver. The frequency response of this pulse is presented in Figure 5.29(b). The sampling frequency indicated by the radar, once the initialization sequence has been completed, indicates a value of 36.04 Gs/s. According to the frequency response of the pulse recorded by the radar, we can note a bandwidth at −3 dB to 4.5 GHz between and 6 GHz. For 7 and 8 GHz frequencies, the standardized Digital Signal Processor (DSP) is, respectively, −10 to −20 dB, which will limit the detection ranges at these frequencies. The same pulse recorded on the wideband oscilloscope shows a bandwidth at −3 dB between 2.4 and 6.8 GHz.

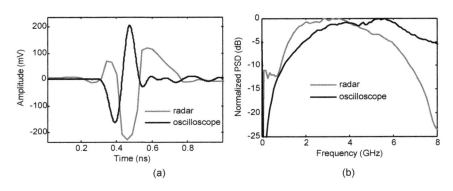

(a)   (b)

Figure 5.29. *a) Pulse rate at the output of the Novelda radar recorded by its own receiver and by the wideband oscilloscope; b) Standardized power spectral density of two signals. For a color version of this figure, see www.iste.co.uk/vena/chipless.zip*

The internal reception stage of the radar, which is mainly composed of an amplifier, promotes the lower frequencies. However, it seems even possible to detect tags whose bandwidth is between 3 and 7 GHz. To perform measurements in open space as those presented earlier, we have added a low-noise amplifier between the receiving antenna and the radar. An averaging operation on 100 measurements is necessary to increase the signal-to-noise ratio of the radar. The temporal calibration procedure used with the temporal measuring bench is repeated in this case. The results obtained for the depolarizing tags based on a double inverted-L are presented in Figure 5.30.

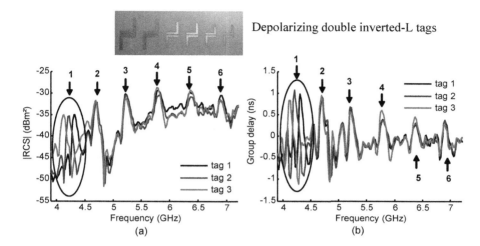

Figure 5.30. *a) Measured RCS of double inverted-L depolarizing tags with the Novelda radar; b) same thing for the group delay. For a color version of this figure, see www.iste.co.uk/vena/chipless.zip*

Figure 5.31 presents the measurement results obtained for the depolarizing tags based on dipoles in short-circuit. In both cases, we clearly note the presence of resonant modes marked in the figures with the numbers 1–6 for the double inverted-L tags and with the numbers 1–8 for the dipole tags in short-circuit. It is interesting to note that the group delay obtained in both cases in Figures 5.30(b) and (b) reflects a redundancy of information in relation to the amplitude variation. This can provide a certain robustness at the detection. The usable frequency range is from a value below 3 GHz up to more than 7 GHz, which is perfectly suitable for the detection of these tags. In accordance with the configurations of the measured tags, only mode No. 1 is modified, as we can actually verify on these curves. The other modes are not modified, the curves are therefore superimposed as expected.

In order to better understand the performance of this radar, we have superimposed the measurement results obtained with the frequency measuring bench and the radar. In Figure 5.32, we can compare the measurements performed, respectively, on the depolarizing double inverted-L tags, numbered 1 and 3. We observe that resonance peaks are placed at the right frequencies and that the overall waveform obtained with the radar approaches the rate waveform with the frequency bench. We note that the peak apex is slightly less marked with the radar, but in all cases, the peaks are clearly recognizable.

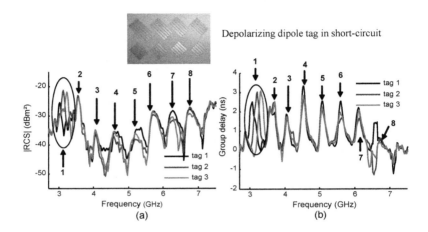

Figure 5.31. *a) Measured RCS of depolarizing dipole tags in short-circuit with the Novelda radar; b) same thing for the group delay. For a color version of this figure, see www.iste.co.uk/vena/chipless.zip*

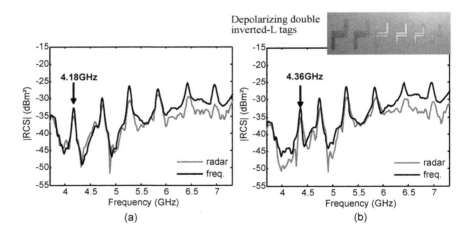

Figure 5.32. *a) Comparison of measurements of the double inverted-L tag, obtained with the frequency bench and the Novelda radar: a) for tag No. 1 and b) for tag No. 3. For a color version of this figure, see www.iste.co.uk/vena/chipless.zip*

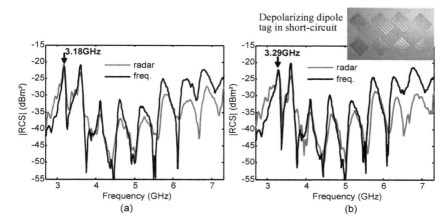

Figure 5.33. *a) Comparison of measurements of the depolarizing dipole tag in short-circuit, obtained with the frequency bench and the Novelda radar: a) for tag No. 2 and b) for tag No. 3. For a color version of this figure, see www.iste.co.uk/vena/chipless.zip*

We have also obtained very good results with the depolarizing dipole tags in short-circuit whose configurations 2 and 3 are presented in Figure 5.33. As explained in Chapter 4, in this case, the peaks located at the highest frequencies are less selective than for double inverted-L tags. This lower selectivity seems to reduce the differences between the frequency measurements and the radar which appeared for the double inverted-L tags. Finally, we keep an almost similar waveform between the two approaches, up to 7 GHz.

Through the implementation of a UWB pulse radar, which is normally intended for localization applications, we have shown that it is possible to detect the electromagnetic signature of chipless tags with the implementation of filtering and calibration procedures. Due to the compactness of the radar and its relatively reduced cost (approximately 2,000 € with the low-noise amplifier), these results constitute the proof of concept of an efficient, robust and inexpensive chipless RFID system.

5.3.5. *Signal formatting and decoding*

Now that we are able to record the radar signature of a chipless tag in practice with frequency and temporal measurement benches and a radar, we must link this signature to a particular ID. For an identification application, we can consider to record the electromagnetic responses of all tags and store them in a database. This database will be consulted at a later stage at each new tag detection, to verify if the electromagnetic signature is similar to an entry in the database. However, this system has limits, especially when the number of tags increases significantly. In fact, the comparison time is proportional to the number of entries in the database, which can lead to incompatible reading times for a given application.

The second possible solution proves to be more efficient for a mass identification application. In this case, we have to detect the frequency position of each resonant mode since we use a frequency PPM coding. We can then deduce the associated code (or digit) at each mode as a function of its frequency gap in relation to the minimum frequency of its variation range and the chosen frequency resolution, as it is shown by equations [4.3]–[4.6] introduced in Chapter 4. The resonant mode detection requires the detection of a peak in the spectrum. In practice, we have observed that it is more robust to detect a dip/peak/dip sequence, in order to be sure that the resonance peak corresponds to a resonant mode. This allows us to sort the valid peaks as a function of their amplitudes compared to the neighboring dips. Regarding the signal processing, a time gating and an envelope filtering in the form of a low-pass filter or a moving average allows us in most cases to overcome the extremely fast amplitude variations in the spectrum. This enables us to limit the false resonant mode detections during the decoding phase.

5.3.6. *Measurements in a real environment*

To complete this chapter, and in order to handle all practical issues [VEN 14] which must be taken into account to develop a detection system of chipless RFID tags in a comprehensive manner, we are interested here in certain tag measurements in an actual environment. Unlike an anechoic environment, in a real environment, the effects listed below will intervene and, most often, will degrade the measurement:

– the reflections of the surrounding objects which can hide the tag response;

– the multi-paths;

– the wireless communications in the UHF (900 MHz and 2.45 GHz) and SHF (5.8 GHz) bands, particularly with the GSM network, and the omnipresent Wi-Fi network in the buildings.

We consider here that the surrounding objects are not mobile. Thus, the reflection/multi-path "mapping" does not vary in time. In the same way as in an anechoic chamber, a measurement in advance of the environment without a tag will allow us to have the overall information on this mapping. The multi-paths are linked to the reflections of the tag response on the neighboring objects and will be added to the first echo received by the detection system. From a practical point of view, their effect will tend to discretize the frequency response of the tag, and therefore to degrade its waveform. To mitigate this effect, a time gating can be used to keep only the first received echo. Finally, the wireless communications, which are omnipresent in our environment, can interfere with certain tag frequencies, particularly at 2.45 and 5.8 GHz. These communications vary as a function of time and are therefore uncorrelated with the tag response that is the same regardless of the query time. By applying an averaging technique on several measurements, in the same way as for the white noise, the effect of these wireless communications will therefore be mitigated.

We have used the frequency measuring bench, which was introduced earlier, to measure the depolarizing tags described in Chapter 4, and whose measurements in an anechoic environment are presented in Figures 5.30 and 5.31. Photos of the measurement environment only for the tag, and for the tag placed on a cardboard box, are presented, respectively, in Figures 5.34(a) and (b). The room used for the measurement contains tables, chairs and many metallic elements. The antenna used is a dual-polarized SATIMO QH 2000 (Open Boundary wideband Quad Ridge antenna), which operates in the 2–32 GHz band. Its gain varies between 6 and 11 dBi between 2 and 10 GHz. We therefore use a monostatic radar configuration, because the source and the receiver are placed in the same location. Port 1 of the VNA is connected via the antenna port, allowing for vertical polarization (connector at the underside), while port 2 is connected via the antenna port, which allows for horizontal polarization (connector on the side). Parameter S_{21} at the VNA level allows us to obtain the tag cross-polarized response

VH (Vertical – Horizontal), while the S_{11} parameter allows us to measure the copolarized response VV (Vertical – Vertical).

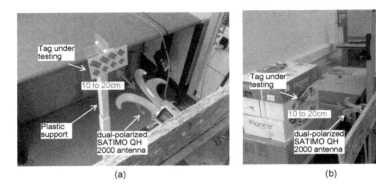

(a)   (b)

Figure 5.34. *a) Photo of a depolarizing dipole tag in short-circuit at 10 cm of the measurement antenna; b) photo of a double "C" tag placed on a cardboard box filled with sheets of paper. In both cases, the antenna used is a dual-polarized antenna. Its operating frequency is between 2 and 32 GHz*

As we can see in Figure 5.34(b), the antenna is oriented toward a plaster wall that separates the measurement part from another desk. The first measurements have been performed on tags placed on a plastic support, whose influence on the electromagnetic response is very low. The results obtained for the depolarizing tags are presented in Figure 5.35 for a separation of 10 cm with the reading antenna, and a transmitting power of 0 dBm. These curves have been obtained by the subtraction of the tag measurement with the measurement in an empty environment (without a tag). We thus observe that a calibration procedure using a reference object as described above is not necessary to find the tag ID. The necessary condition so as not to perform the reference object measurement is that the reception stage of the detection system does not modify the shape of the reflected signal considerably. However, when we want to detect very small amplitude or phase variations on the reflected signal, the calibration procedure described in section 5.3.1.1 is mandatory. An averaging on 10 measurements has been implemented, in order to increase the signal-to-noise ratio in reception, which is not necessary in an anechoic environment. The curves presented in Figures 5.35(a) and (b), respectively, represent the response of depolarizing dipole tags in short-circuit and the response of double inverted-L tags for three different configurations each time. The peaks can be easily distinguished, and the resonance frequency differences related to parameter variations of each tag can be detected without error. In this case, we do not observe any degradation compared to an anechoic environment.

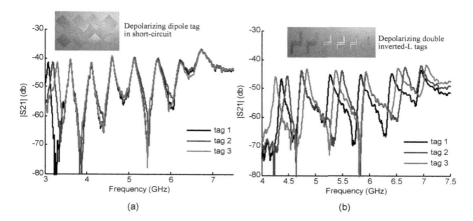

Figure 5.35. *Cross-polarization measurements of depolarizing tags: a) dipole in short-circuits; b) double inverted-L. The tags are placed on a not very intrusive plastic support as presented in Figure 5.34(a), the support response remains low. The presented results have been obtained simply by subtracting the measurement of tags with the empty environment measurement. The dimensions of the tags are provided in Chapter 4. For a color version of this figure, see www.iste.co.uk/vena/chipless.zip*

If we are interested in double "C" tags, which respond only in the copolarization of the incident wave, good results can also be obtained as shown in Figure 5.36(a). It should be noted, however, that the response is extracted from the subtraction of the S_{11} parameter measured in the presence of a tag and without a tag, and especially that the support which is used to place the tag is not very reflective and slightly disrupts the tag signature. The signals that we are trying to measure here are well below the reflection level related to the simple antenna mismatching, as shown in Figure 5.36(b), or the tag signature is not visible. We obtain, however, correct measurements.

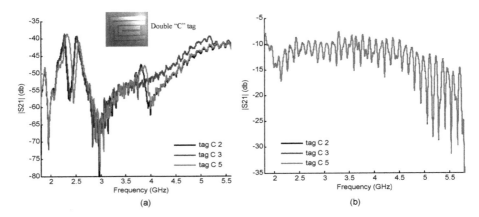

Figure 5.36. *a) Measurements of double "C" tags placed on a not very intrusive plastic support for the copolarization measurement (see Figure 5.34(a)). The curves presented are the measurements of different tags subtracted at the empty environment measurement. The dimensions of the tags are provided in Chapter 4; b) direct measurement of the tag without subtraction with the empty environment measurement. It shows mainly the antenna reflection level on its excitation port in vertical polarization. For a color version of this figure, see www.iste.co.uk/ vena/chipless.zip*

To approach an even more practical application, we have also performed measurements of tags, this time on a cardboard box filled with sheets of paper, as shown in Figure 5.34(b). We are trying to identify a large-size object compared to the tag. Concretely for the measurement, the object reflection is much greater than that of the tag and will therefore greatly disturb the measurement if no action takes place to take it into account [VEN 13a]. Figure 5.37 shows the influence of the reading distance on the response of the depolarizing dipole tag in short-circuits, i.e. the tags which are specially designed to address this problem directly related to the identification of objects.

In the simplest case (which, unfortunately, is not representative), the cardboard box to identify has a known geometry, and it is always placed at the same location, i.e. at a known distance. Therefore, we can measure the response of the environment containing the cardboard box and subtract it from the tag response, which provides the curves in Figure 5.37(a). We can observe that, regardless of the distance of the tag, the resonance peaks are in all distinguishable without interpretation errors.

In a second case, we will seek to cover the most realistic configuration, i.e. the cardboard box has variables dimensions and its positioning is also unknown. We do not know its response in advance. It is, therefore, better to simply subtract the response of the empty environment, i.e. without the cardboard box, at the response of the tag placed on the cardboard box. The results obtained in this case are presented in Figure 5.37(b). Again the results obtained are very satisfactory, as we can extract all resonance peaks at a reading distance of up to 20 cm. The empty environment measurement does not involve the cardboard box. A single and unique measurement to characterize the environment has therefore been carried out in addition to the three tag measurements on the cardboard box corresponding to the three different distances. We record four measurements in total.

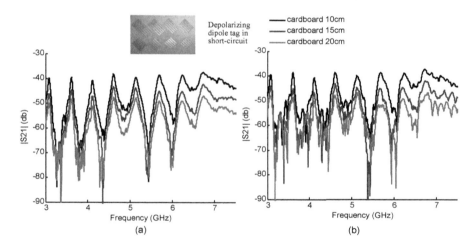

Figure 5.37. *Cross-polarization measurements of the depolarizing dipole tag in short-circuit No. 1 placed on an object with a large RCS, namely, a cardboard box filled with sheets of paper (Figure 5.34(b)): a) tag measurements subtracted from the response carried out in the presence of the cardboard box. Measurements have been carried out for different distances; b) tag measurements subtracted from the response of the empty environment, i.e. without the cardboard box. The dimensions of the tag are provided in Chapter 4. For a color version of this figure, see www.iste.co.uk/vena/chipless.zip*

The results presented in Figure 5.37(b) show the reading robustness when depolarizing tags are used. We note that the tag response is generated in both the vertical and horizontal polarization, while the tag excitation is carried out in the vertical polarization. A theoretical model of the problem is presented in [VEN 13a]. A common object such as a cardboard box does not depolarize. It therefore tends to reflect the wave only in copolarization, i.e. in the vertical polarization. It is therefore "transparent" from the point of view of the cross-polarization (horizontal polarization, in this case) and will not interfere with the tag response in this polarization. That is exactly what we observe if we compare Figures 37(a) and (b), where there are very few significant differences.

Figures 5.38(a) and (b) present, respectively, the response of the cardboard box filled with sheets in cross-polarization and in copolarization confirm this behavior. We notice a level of an almost constant response of −55 dB in cross-polarization, close to the antenna isolation level, while a variable response level that is close to −25 dB can be measured in copolarization. This provides a gap of 30 dB between the two polarizations.

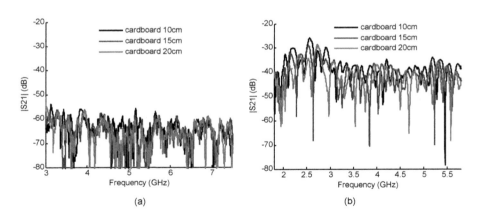

(a)   (b)

Figure 5.38. *Measurements of the response of the cardboard box at distances of 10, 15 and 20 cm subtracted from the response of the empty environment (without a cardboard box): a) in cross-polarization and b) in copolarization. For a color version of this figure, see www.iste.co.uk/vena/chipless.zip*

Due to the large and variable response of the cardboard box in the copolarization, a tag such as the double "C" tag can be used only in a very particular case, namely, when the cardboard box has a precisely known geometry and positioning. This is the only condition where it is possible to characterize its electromagnetic response. Thus, Figure 5.39(a) shows the measurement results of the double "C" tag No. 2 in copolarization for several distances. The tag response has been recorded for each distance, in order to be able to subtract it from the response of the tag/cardboard box assembly (it has also been obtained for each distance, or six measurements in total). The curves obtained are more noisy than those relating to the depolarizing tags, but the resonance peaks can still be distinguished. We have also reproduced in a very exact manner the case where we consider the object as known. If we now consider that the object is unknown, the results will change radically. In fact, this second measurement scenario where the response of the cardboard box is not known in advance (i.e. here it is not measured separately) is very different from the point of view of measurement. Figure 5.39(b) presents these results, i.e. when the empty environment measurement, without a cardboard box, is subtracted from the response of the tag/cardboard box assembly for each of the three distances (four measurements in total). In this case, we can see that it is impossible to distinguish the tag ID.

These measurements, which have been carried out in a real environment, provide the evidence that it is quite possible to identify chipless tags in the presence of various disturbance sources, particularly those that we find typically in buildings. Here, we prove that the detection of a tag on an object such as a cardboard box filled with sheets of paper is possible. In addition, through measurements carried out by varying the reading distance between the antenna and the tag placed on the cardboard box, we have shown that a depolarizing tag brings a much higher detection robustness. It proves to be discriminant in practice to the extent that it allows us to overcome a large number of measurements which could not be implemented in an operational context. In fact, in other words, this principle allows the decoding of the tag ID without knowing the object on which the tag is placed in advance. Furthermore, a positioning variation of approximately 20 cm in no way alters the resonances peaks. We have seen previously that the measurement results obtained with the temporal measuring bench and the Novelda radar are very close to the measurements obtained in frequency. We can conclude that it is quite possible to achieve similar results in a real environment with impulse radio detection systems [GAR 15a, GAR 15b].

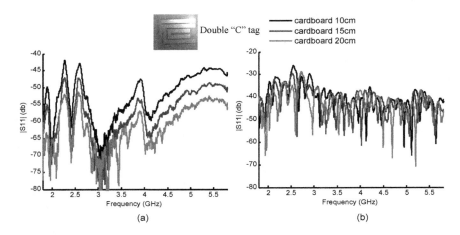

Figure 5.39. *Copolarization measurements of the double "C" tag No. 2 placed on a cardboard box filled with sheets of paper, highly reflective object (Figure 5.34(b)): a) tag measurements subtracted from the measured response of the cardboard box, respectively, at distances of 10, 15 and 20 cm; b) tag measurements subtracted from the response of the empty environment, without a cardboard box (a single measurement for the three distances). The dimensions of the tag are provided in Chapter 4. For a color version of this figure, see www.iste.co.uk/vena/chipless.zip*

5.4. Conclusion

In this chapter, we have analyzed one of the fundamental aspects of the development of chipless RFID technology, namely, the tag implementation and measurement. We have seen that the major interest of chipless RFID tags comes from the fact that they can be manufactured with the use of mass printing processes, such as flexography, or, for smaller quantities, with conductive ink jet printing processes. The substrates that can be used are paper or flexible plastic substrates, which allows us to achieve unit manufacturing costs lower than a euro cent compared to minimum 10–30 cents of the conventional passive RFID tags. Due to the fact that this technology is relatively recent and is based on quite a different coding principle from the conventional modulation schemes, today there are no off-the-shelf readers. In contrast, we have demonstrated that with a frequency or temporal bistatic radar measuring bench, we are able to extract the electromagnetic signature of a chipless tag with a calibration procedure of three measurements. In order to meet the UWB standards which define the very restrictive transmission masks for the ultra wideband communicating systems, the design of an impulse radio-based radar detection system seems more judicious. In fact, in the case of signals with low duty cycles, we are able to concentrate a relatively high energy in a very short time, so as to comply with the transmission masks which concern the average radiated powers. We have, therefore, shown that with a commercially available impulse radar, which is normally dedicated to localization applications, it is possible to extract the electromagnetic signatures of chipless tags while approaching significantly the results obtained with the frequency measuring bench. Measurements carried out outside of an anechoic chamber, in a real environment, showed that it was possible to detect the ID of a tag placed on a cardboard box filled with sheets of paper, with a greater measurement time, to improve the signal-to-noise ratio. The depolarizing tags have shown a potential inequality to the extent that they have allowed us to detect the signature of a tag placed on a cardboard box without knowing in advance its electromagnetic response. This enables us, among others, to detect the tags on objects of variable geometries, whose place opposite of the antenna can vary in a range of approximately 20 cm. This result is remarkable to the extent that, by considerably limiting the reading constraints, this technology becomes compatible with most of the requirements that we find in the identification sectors. Finally, paper tags have been created by a conductive ink printing process, with a coding capacity of approximately 19 bits, which can be read by a commercially available pulse radar respecting the UWB standards, all in a real environment. These studies thus show the potential of chipless RFID technology, which should be very important in the identification field in the years to come. Similarly, to conclude on an even more positive note concerning the application potential of chipless technology, we note in [GAR 15a] and [GAR 15b] that a chipless reader which is compatible with the regulations and with a unit cost of approximately 1,000 € has been developed and shows very good reading results.

Conclusion

In this book, we have analyzed the latest developments on radio frequency identification (RFID) technology in its entirety in Chapter 1, before addressing the latest developments of chipless RFID technologies in Chapter 2. This has allowed us to place chipless tags, according to performance criteria and their cost, among the other RFID technologies. This has also allowed us to define the different issues to address, to make chipless RFID tags more efficient. First, we have identified that the size of the tags should be reduced, and that coding densities should be improved. New techniques to significantly increase the detection robustness of tags in a real environment have been presented. The problem concerning the design of a reading system has been addressed in Chapter 5.

Chapter 3 has allowed us to formalize coding designs that can be used in a chipless tag whose information is coded in particular in the frequency domain. We have introduced performance criteria on coding, such as surface density coding and spectral density coding, to better evaluate the coding of a given tag. The frequency-based pulse position modulation (PPM) coding that we have adopted in almost all of our designs shows a better effectiveness than the coding in the absence/presence of peaks. The hybrid coding design has notably been introduced to increase the number of possible combinations for a given resonator. This design has been successfully transposed to the design of tags based on the combination of "C" resonators.

In Chapter 4, we introduced the design approach of RF encoding particles (REP) chipless tags. Different designs developed to address various problems mentioned above have been presented. In order to reduce the dimensions of the tags, we have used multiple resonators that act as a receiving antenna, a filter and a transmitting antenna. The "C" resonators, which were used, are reduced in size and have a quarter-wavelength resonance mode, which reduces their maximum dimension by two, compared to a dipole in short-circuit. They also have the advantage of not requiring a ground plane. To make their use possible even when they are placed on objects which will induce a shift on their resonance frequency, we have developed a compensation technique based on the use of one or several resonators whose role is to detect the effective permittivity of the surrounding environment of the tag. Next, we have presented tags that are based on the use of circular patch resonators. With this structure, we have improved detection robustness, as its electromagnetic response is the same regardless of its orientation. In addition, by being able to insert the circles within each other and side-by-side without there being too much coupling has allowed us to obtain a coding of 49 bits on a surface of 4×4 cm^2. By slightly modifying circular resonators with the presence of an aperture, we have transformed a polarization independent structure in a structure which is very sensitive to the orientation. This has allowed us to explore a new coding design by taking advantage of the polarization diversity of the reading system. The advantage of this design is essentially to limit the required bandwidth in such a way that only the industrial, scientific, medical (ISM) bands can be used, while retaining a significant coding capacity. Finally, the last design presented in this chapter concerned tags that are able to depolarize the incident wave, or in other words, to generate the tag response on an orthogonal polarization. This allows us to isolate the transmission signals from the reception signals, providing in this way an improved sensitivity in reception. With this approach, we can also improve detection robustness in disturbed environments by echo-generating objects. In fact, everyday objects do not tend to depolarize an incidental electromagnetic wave, unlike depolarizing tags which are optimized for this task.

Finally, we presented studies that have enabled us to provide the proof of concept of the application potential of the created designs. First, we have been able to see that chipless RFID tags can be created by printing processes on low-cost substrates such as paper. This has allowed us to achieve unit manufacturing costs lower than 0.5 euro cents. The measuring benches used to detect the spectral responses of the tags have been described in detail. A frequency bistatic radar measuring bench has enabled us to obtain very reliable measurements in simulations. For the purpose of designing a detection system that can respond to the transmission masks defined by the federal communications commission (FCC) and the electronic communications committee (ECC), we verified that the use of a time domain reading system may provide great performances. For this, we have shown that the use of a commercially available impulse radio-based ultra wide band (UWB) radar, which is typically intended for tracking applications of objects or people, makes possible the detection of a chipless tag signature on a bandwidth of 3–7 GHz.

Bibliography

[ARC 11] ARCEP, Observatoire des activités postales 2010, 2011.

[BAL 05] BALANIS C., *Antenna Theory: Analysis and Design*, Wiley, 2005.

[BAL 09a] BALBIN I., KARMAKAR N., "Novel chipless RFID tag for conveyor belt tracking using multi-resonant dipole antenna", *European Microwave Conference*, pp. 1109–1112, 2009.

[BAL 09b] BALBIN I., KARMAKAR N., "Phase-encoded chipless RFID transponder for large-scale low-cost applications", *IEEE Microwave and Wireless Components Letters*, vol. 19, no. 8, pp. 509–511, 2009.

[BER 11] BERNIER M., GARET F., PERRET E. *et al.*, "Terahertz encoding approach for secured chipless radio frequency identification", *Applied Optics*, vol. 50, no. 23, pp. 4648–4655, 2011.

[BLI 09] BLISCHAK A., MANTEGHI M., "Pole residue techniques for chipless RFID detection", *IEEE Antennas and PROPAGATION Society International Symposium*, pp. 1–4, 2009.

[BRA 24] BRARD E., Process for radiotelegraphic or radiotelephonic communication, French Pat. 1924, U.S. Pat. 1744036, 1930.

[BRO 13] BROOKER G., GOMEZ J., "Lev Termen's Great Seal bug analyzed", *IEEE Aerospace and Electronic Systems Magazine*, vol. 28, pp. 4–11, 2013.

[BRO 99] BROWN L., *Technical and Military Imperatives: A Radar History of World War 2*, CRC Press, 1999.

[CHA 06] CHAMARTI A., VARAHRAMYAN K., "Transmission delay line based ID generation circuit for RFID applications", *IEEE Microwave and Wireless Components Letters*, vol. 16, no. 11, pp. 588–590, 2006.

[CHE 82] CHEW W., "A broad-band annular-ring microstrip antenna", *IEEE Transactions on Antennas and Propagation*, vol. 30, pp. 918–922, 1982.

[COS 13] COSTA F., GENOVESI S., MONORCHIO A., "A chipless RFID based on multiresonant high-impedance surfaces", *IEEE Transactions on Microwave Theory and Techniques*, vol. 61, pp. 146–153, 2013.

[COS 14] COSTA F., GENOVESI S., MONORCHIO A. *et al.*, "Calibration method for periodic surface based chipless tags", *IEEE RFID Technology and Applications Conference (RFID-TA)*, pp. 78–81, 2014.

[DAR 08] DARDARI D., D'ERRICO R., "Passive ultrawide bandwidth RFID", *IEEE Global Telecommunications Conference*, New Orleans, USA, pp. 1–6, 2008.

[DAR 10] DARDARI D., D'ERRICO R., ROBLIN C. *et al.*, "Ultrawide bandwidth RFID: the next generation?", *Proceedings of the IEEE*, vol. 98, no. 9, pp. 1570–1582, 2010.

[DEA 05] DEARDEN A.L., SMITH P.J., SHIN D. *et al.*, "A low curing temperature silver ink for use in ink-jet printing and subsequent production of conductive tracks", *Macromolecular Rapid Communications*, vol. 26, no. 4, pp. 315–318, 2005.

[DEC 16] DECAWAVE, available at: www.decawave.com, 2016.

[DEE 10] DEEPU V., VENA A., PERRET E. *et al.*, "New RF identification technology for secure applications", *IEEE International Conference on RFID-Technology and Applications (RFID-TA)*, pp. 159–163, 2010.

[DEN 09] DENNEULIN A., BRAS J., BLAYO A. *et al.*, "The influence of carbon nanotubes in inkjet printing of conductive polymer suspensions", *Nanotechnology*, vol. 20, p. 385701, 2009.

[DIM 16] DIMATIX MATERIALS, available at http://www.fujifilmusa.com/ products/industrial_inkjet_ printheads / deposition-products / index.html, 2016.

[ETS 10] ETSI E. 302 065 V1.2.1, "Electromagnetic compatibility and radio spectrum matters (ERM); ultra wideband (UWB) technologies for communication purposes; harmonized en covering the essential requirements of article 3.2 of the RTTE directive", *European Standard Telecommunication Institute (ETSI)*, 2010.

[FIN 10] FINKENZELLER K., *RFID Handbook: Fundamentals and Applications in Contactless Smart Cards, Radio Frequency Identification and Near-Field Communication*, Wiley, 2010.

[FLE 02] FLETCHER R.R., Low-cost electromagnetic tagging: design and implementation, PhD Thesis, available at: cba.mit.edu/docs/theses/02.09. fletcher.pdf, 2002.

[GAR 15a] GARBATI M., SIRAGUSA R., PERRET VENA A., *et al.*, "High performance chipless RFID reader based on IR- E. UWB technology", *9th European Conference on Antennas and Propagation (EuCAP'2015)*, Lisbon, Portugal, pp. 1–5, 2015.

[GAR 15b] GARBATI M., SIRAGUSA R., PERRET E. *et al.*, "Low cost low sampling noise UWB Chipless RFID reader", *IEEE MTT-S International in Microwave Symposium (IMS 2015)*, Phoenix, USA, pp. 1–4, 2015.

[GIR 12] GIRBAU D., LAZARO A., RAMOS A., "Time-coded chipless RFID tags: design, characterization and application", *IEEE International Conference on RFID-Technologies and Applications (RFID-TA)*, pp. 12–17, 2012.

[GLI 00] GLINSKY A., *Theremin: Ether Music and Espionage*, University of Illinois Press, 2000.

[GRE 63] GREEN R.B., The general theory of antenna scattering, OSU Report, No. 1223-17. 1963.

[GUI 10] GUIDI F., DARDARI D., ROBLIN C. *et al.*, "Backscatter communication using ultrawide bandwidth signals for RFID applications", in GIUSTO D., IERA A., MORABITO G. *et al.* (eds), *The Internet of Things: 20th Tyrrhenian Workshop on Digital Communications*, Springer, pp. 251–261, 2010.

[GUP 10a] GUPTA S., NIKFAL B., CALOZ C., "RFID system based on pulse-position modulation using group delay engineered microwave C-sections", *Asia-Pacific Microwave Conference (APMC)*, Yokohama, Japan, pp. 203–206, 2010.

[GUP 10b] GUPTA S., PARSA A., PERRET E. *et al.*, "Group-delay engineered noncommensurate transmission line all-pass network for analog signal processing", *IEEE Transactions on Microwave Theory and Techniques*, vol. 58, no. 9, pp. 2392–2407, 2010.

[HAM 11] HAMDI M., GARET F., DUVILLARET L. *et al.*, "THID Tags for identification in the THz domain", *6èmes Journées Térahertz*, La Grande Motte, France, 2011.

[HAM 12] HAMDI M., GARET F., DUVILLARET L. *et al.*, "New approach for chipless and low cost identification tag in the THz frequency domain", *IEEE International Conference on RFID-Technology and Applications*, Nice, France, pp. 24–28, 2012.

[HAM 13] HAMDI M., GARET F., DUVILLARET L. *et al.*, "Identification tag in the THz frequency domain using low cost and tunable refractive index materials", *Annals of Telecommunications, Special Issue on Chipless RFID*, vol. 68, pp. 415–424, 2013.

[HAR 02] HARTMANN C., "A global saw ID tag with large data capacity", *IEEE Ultrasonics Symposium Proceedings*, pp. 65–69, 2002.

[HAR 04] HARTMANN C., HARTMANN P., BROWN P. *et al.*, "Anti-collision methods for global saw RFID tag systems", *IEEE Ultrasonics Symposium,* pp. 805–808, 2004.

[HAR 09] HARROP P., DAS R., *Chipless RFID Forecasts, Technologies & Players 2009-2019*, IDTechEx, 2009.

[HAR 52] HARRIS D., Radio transmission systems with modulatable passive responder, U.S. Pat. 2927321, filed in 1952.

[HAR 63] HARRINGTON R., "Electromagnetic scattering by antennas", *IEEE Transactions on Antennas and Propagation*, vol. 11, no. 5, pp. 595–596, 1963.

[HAR 64] HARRINGTON R., "Theory of loaded scatterers", *Proceedings of the Institution of Electrical Engineers*, vol. 111, no. 4, pp. 617–623, 1964.

[HOP 08] HOPKINS R., FREE C., "Equivalent circuit for the microstrip ring resonator suitable for broadband materials characterisation", *IET Microwaves, Antennas & Propagation*, vol. 2, pp. 66–73, 2008.

[HUA 11] HUANG L., HUANG Y., LIANG J. *et al.*, "Graphene-based conducting inks for direct inkjet printing of flexible conductive patterns and their applications in electric circuits and chemical sensors", *Nano Research*, vol. 4, no. 7, pp. 675–684, 2011.

[IDT 16] IDTECHEX, available at http://www.idtechex.com/, 2016.

[IMP 16] IMPINJ, available at: www.impinj.com, 2016.

[INK 16] INKSURE TECHNOLOGIES, available at: www.inksure.com, 2016.

[JAL 05a] JALALY I., ROBERTSON I., "Capacitively-tuned split microstrip resonators for RFID barcodes", *European Microwave Conference*, pp. 1–4, 2005.

[JAL 05b] JALALY I., ROBERTSON I., "RF barcodes using multiple frequency bands", *IEEE MTT-S International Microwave Symposium Digest*, pp. 139–142, 2005.

[JAN 10] JANG H., LIM W., OH K. *et al.*, "Design of low-cost chipless system using printable chipless tag with electromagnetic code", *IEEE Microwave and Wireless Components Letters*, vol. 20, no. 11, pp. 640–642, 2010.

[KAR 10] KARMAKAR N., *Handbook of Smart Antennas for RFID Systems*, Wiley Online Library, 2010.

[KIM 08] KIM J., Saw based chipless passive RFID tag using cellulose paper as substrate and method for manufacturing the cellulose paper, WO Pat. WO/2008/056,848, 2008

[KIR 07] KIRIAZI J., NAKAKURA J., LUBECKE V. *et al.*, "Low profile harmonic radar transponder for tracking small endangered species", *29th Annual International Conference of the IEEE Engineering in Medicine and Biology Society*, pp. 2338–2341, 2007.

[KON 09] KONSTAS Z., RIDA A., VYAS R. *et al.*, "A novel green inkjet-printed z-shaped monopole antenna for RFID applications", *3rd European Conference on Antennas and Propagation*, pp. 2340–2343, 2009.

[LAN 05] LANDT J., "The history of RFID", *Potentials*, IEEE, vol. 24, no. 4, pp. 8–11, 2005.

[LAZ 11] LAZARO A., RAMOS A., GIRBAU D. *et al.*, "Chipless UWB RFID tag detection using continuous wavelet transform", *IEEE Antennas and Wireless Propagation Letters*, vol. 10, pp. 520–523, 2011.

[LEE 05] LEE J., KIM H., "Radar target discrimination using transient response reconstruction", *Journal of Electromagnetic Waves and Applications*, vol. 19, no. 5, pp. 655–669, 2005.

[MAN 09] MANDEL C., SCHUSSLER M., MAASCH M. *et al.*, "A novel passive phase modulator based on LH delay lines for chipless microwave RFID applications", *IEEE MTT-S International Microwave Workshop on Wireless Sensing, Local Positioning, and RFID*, pp. 1–4, 2009.

[MCV 06] MCVAY J., HOORFAR A., ENGHETA N., "Theory and experiments on Peano and Hilbert curve RFID tags", *Proceedings of SPIE*, p. 624808, 2006.

[MEN 95] MENG B., BOOSKE J., COOPER R., "Extended cavity perturbation technique to determine the complex permittivity of dielectric materials", *IEEE Transactions on Microwave Theory and Techniques*, vol. 43, pp. 2633–2636, 1995.

[MOO 07] MOORHOUSE C., VILLARREAL F., BAKER H. *et al.*, "Laser drilling of copper foils for electronics applications", *IEEE Transactions on Components and Packaging Technologies*, vol. 30, no. 2, pp. 254–263, 2007.

[MUC 08] MUCHKAEV A., Carrierless RFID system, Pat. WO/2007/123849, 2008.

[MUK 07a] MUKHERJEE S., "Chipless radio frequency identification by remote measurement of complex impedance", *European Conference on Wireless Technologies*, pp. 249–252, 2007.

[MUK 07b] MUKHERJEE S., "Chipless radio frequency identification (RFID) device", *1st Annual RFID Eurasia*, Istanbul, Turkey, pp. 1–4, 2007.

[MUK 09] MUKHERJEE S., CHAKRABORTY G., "Chipless RFID using stacked multilayer patches", *Applied Electromagnetics Conference*, pp. 1–4, 2009.

[NAI 13a] NAIR R.S., PERRET E., TEDJINI S., "A temporal multi-frequency encoding technique for chipless RFID based on C-sections", *Progress In Electromagnetics Research B*, vol. 49, pp. 107–127, 2013.

[NAI 13b] NAIR R.S., PERRET E., TEDJINI S., "Group delay modulation for pulse position coding based on periodically coupled C-sections", *Annals of Telecommunications, Special Issue on Chipless RFID*, vol. 68, nos. 7–8, pp. 447–457, 2013.

[NAI 11] NAIR R., PERRET E., TEDJINI S., "Chipless RFID based on group delay encoding", *IEEE International Conference on RFID-Technologies and Applications*, pp. 214–218, 2011.

[NIK 06] NIKITIN P., RAO K., "Performance limitations of passive UHF RFID systems", *Proceedings of the IEEE Antennas and Propagation Symposium, Albuquerque*, USA, pp.1011–1014, 2006.

[NIK 12] NIKITIN P., "Leon Theremin (Lev Termen)", *IEEE Antennas and Propagation Magazine*, vol. 54, pp. 252–257, 2012.

[NOV 16] NOVELDA, available at http://www.novelda.no/, 2016.

[NXP 16] NXP, available at: www.nxp.com, 2016.

[NYS 88] NYSEN P., SKEIE H., ARMSTRONG D., System for interrogating a passive transponder carrying phase-encoded information, US Patent 4725841, 1988.

[PAR 09] PARET D., *RFID at Ultra and Super High Frequencies: Theory and Application*, Wiley, 2009.

[PER 11a] PERRET E., HAMDI M., VENA A. *et al.*, "RF and THz identification using a new generation of chipless RFID tags", *Radioengineering – Special Issue: Emerging Materials, Methods, and Technologies in Antenna & Propagation*, vol. 20, no. 2, pp. 380–386, June 2011.

[PER 11b] PERRET E., TEDJINI S., VASUDEVAN NAIR D. *et al.*, "Chipless passive RFID tag", WO Pat. WO/2011/098,719 (A2), French Pat 2956232 (B1), 2011.

[PER 12] PERRET E., TEDJINI S., NAIR R., "Design of antennas for UHF RFID tags", *Proceedings of the IEEE Special Issue on Wireless Communication Antennas*, vol. 100, no. 7, pp. 2330–2340, 2012.

[PER 13] PERRET E., HAMDI M., TOURTOLLET G.E.P. *et al.*, "THID, the next step of chipless RFID", *7th IEEE RFID Conference*, Florida, USA, pp. 261–268, 2013.

[PER 14] PERRET E., NAIR R.S., KAMEL E.B. *et al.*, "Chipless RFID tags for passive wireless sensor grids", *XXXIth URSI General Assembly and Scientific Symposium (URSI GASS)*, Beijing, China, 2014.

[PER 14] PERRET E., *Radio Frequency Identification and Sensors: From RFID to Chipless RFID*, ISTE, London and John Wiley & Sons, New York, 2014.

[PRE 08] PRERADOVIC S., BALBIN I., KARMAKAR N. *et al.*, "A novel chipless RFID system based on planar multiresonators for barcode replacement", *IEEE International Conference on RFID*, pp. 289–296, 2008.

[PRE 09a] PRERADOVIC S., ROY S., KARMAKAR N., "Fully printable multi-bit chipless RFID transponder on flexible laminate", *Asia Pacific Microwave Conference*, pp. 2371–2374, 2009.

[PRE 09b] PRERADOVIC S., KARMAKAR N., "Design of fully printable planar chipless RFID transponder with 35-bit data capacity", *European Microwave Conference*, pp. 13–16, 2009.

[PRE 09c] PRERADOVIC S., KARMAKAR N., "Design of short range chipless RFID reader prototype", *5th International Conference on Intelligent Sensors, Sensor Networks and Information Processing*, pp. 307–312, 2009.

[PRE 10a] PRERADOVIC S., KARMAKAR N., ZENERE M., "UWB chipless tag RFID reader design", *IEEE International Conference on RFID-Technology and Applications*, pp. 257–262, 2010.

[PRE 10b] PRERADOVIC S., KARMAKAR N., "4th generation multiresonator-based chipless RFID tag utilizing spiral EBGS", *European Microwave Conference*, pp. 1746–1749, 2010.

[RAM 12] RAMOS A., GIRBAU D., LAZARO A. *et al.*, "IR-UWB radar system and tag design for time-coded chipless RFID", *6th European Conference on Antennas and Propagation (EUCAP)*, pp. 2491–2494, 2012.

[RAM 16] RAMOS A., LAZARO A., GIRBAU D. *et al.*, *RFID and Wireless Sensors using Ultra-Wideband Technology*, ISTE, London and John Wiley & Sons, New York, 2016.

[RÉS 16] RÉSEAU LASER, available at http://laser.agmat.asso.fr, 2016.

[REU 06] REUNAMÄKI J., Ultra wideband radio frequency identification techniques, Pat. WO/2006/070,237, 2006.

[REZ 14] REZAIESARLAK R., MANTEGHI M., *Chipless RFID: Design Procedure and Detection Techniques*, Springer, 2014.

[RFS 16] RFSAW, available at: www.rfsaw.com, 2016.

[SAV 16] SAVI, available at: www.savi.com, 2016.

[SCH 09] SCHULER M., MANDEL C., MAASCH M. *et al.*, "Phase modulation scheme for chipless RFID-and wireless sensor tags", *Asia Pacific Microwave Conference*, pp. 229–232, 2009.

[SHA 11] SHAKER G., SAFAVI-NAEINI S., SANGARY N. *et al.*, "Inkjet printing of ultra-wideband (UWB) antennas on paper-based substrates", *IEEE Antennas and Wireless Propagation Letters*, vol. 10, pp. 111–114, 2011.

[SHR 09] SHRESTHA S., BALACHANDRAN M., AGARWAL M. *et al.*, "A chipless RFID sensor system for cyber centric monitoring applications", *IEEE Transactions on Microwave Theory and Techniques*, vol. 57, no. 5, pp. 1303–1309, 2009.

[STO 48] STOCKMAN H., "Communication by means of reflected power", *Proceedings of the IRE*, no. 36, pp. 1196–1204, 1948.

[SUB 05] SUBRAMANIAN V., FRECHET J., CHANG P. *et al.*, "Progress toward development of all-printed RFID tags: materials, processes, and devices", *Proceedings of the IEEE*, vol. 93, no. 7, pp. 1330–1338, 2005.

[TAG 16] TAGENT, available at: www.tagent.com, 2016.

[TAG 16] TAGSENSE, available at: www.tagsense.com, 2016.

[TED 10] TEDJINI S., PERRET E., DEEPU V. *et al.*, "Chipless tags for RF and THZ identification", *Proceedings of the 4th European Conference on Antennas and Propagation*, pp. 1–5, 2010.

[TEK 09] TEKTRONIX, Making sense of effective bits in oscilloscope measurements, available at: www.tektronix.com, 2009.

[VEM 07] VEMAGIRI J., CHAMARTI A., AGARWAL M. *et al.*, "Transmission line delay- based radio frequency identification (RFID) tag", *Microwave and Optical Technology Letters*, vol. 49, no. 8, pp. 1900–1904, 2007.

[VEN 09] VENA A., ROUX P., "Near field coupling with small RFID objects", *Proceedings of PIERS*, Moscow, Russia, pp. 535–539, 2009.

[VEN 11] VENA A., PERRET E., TEDJINI S., "Chipless RFID tag using hybrid coding technique", *IEEE Transactions on Microwave Theory and Techniques*, vol. 59, no. 12, pp. 3356–3364, 2011.

[VEN 11a] VENA A., PERRET E., TEDJINI S., "Novel compact RFID chipless tag", *PIERS Proceedings, Marrakesh*, Morroco, pp. 1062–1066, 2011.

[VEN 11b] VENA A., PERRET E., TEDJINI S., "RFID chipless tag based on multiple phase shifters", *IEEE MTT-S International Microwave Symposium Digest*, pp. 1–4, 2011.

[VEN 12a] VENA A., PERRET E., TEDJINI S., "A fully printable chipless RFID tag with detuning correction technique", *IEEE Microwave and Wireless Components Letters*, vol. 22, no. 4, pp. 209–211, 2012.

[VEN 12b] VENA A., PERRET E., TEDJINI S., "A compact chipless RFID tag using polarization diversity for encoding and sensing", *IEEE International Conference on RFID*, pp. 191–197, 2012.

[VEN 12c] VENA A., PERRET E., TEDJINI S., "High capacity chipless RFID tag insensitive to the polarization", *IEEE Transactions on Antennas and Propagation*, vol. 60, no. 10, pp. 4509–4515, 2012.

[VEN 12d] VENA A., PERRET E., TEDJNI S., "Design of compact and auto-compensated single-layer chipless RFID tag", *IEEE Transactions on Microwave Theory and Techniques*, vol. 60, no. 9, pp. 2913–2924, 2012.

[VEN 13] VENA A., PERRET E., TEDJINI S., "Design rules for chipless RFID tags based on multiple scatterers", *Annals of Telecommunications, Special Issue on Chipless RFID*, vol. 68 nos. 7–8, pp. 361–374, 2013.

[VEN 13a] VENA A., PERRET E., TEDJINI S., "A depolarizing chipless RFID tag for robust detection and its FCC compliant UWB reading system", *IEEE Transactions on Microwave Theory and Techniques*, vol. 61, no. 8, pp. 2982–2994, August 2013.

[VEN 13b] VENA A., PERRET E., TEDJINI S. et al., "Design of chipless RFID tags printed on paper by flexography", *IEEE Transactions on Antennas and Propagation*, vol. 61, no. 12, pp. 5868–5877, 2013.

[VEN 14] VENA A., PERRET E., SORLI B. *et al.*, "Study on the detection reliability of chipless RFID systems", *XXXIth URSI General Assembly and Scientific Symposium (URSI GASS)*, Beijing, China, pp. 1–4, 2014.

[VEN 15a] VENA A., PERRET E., SORLI B. *et al.*, "Theoretical study on detection distance for chipless RFID systems according to transmit power regulation standards", *9th European Conference on Antennas and Propagation (EuCAP' 15)*, Lisbon, Portugal, pp. 1–4, 2015.

[VEN 15b] VENA A., SYDÄNHEIMO L., TENTZERIS M.M. *et al.*, "A fully inkjet-printed wireless and chipless sensor for CO_2 and temperature detection", *IEEE Sensors Journal*, vol. 15, no. 1, pp. 89–99, 2015.

[VOG 59] VOGELMAN J., Passive data transmission techniques using radar echoes, U.S. Pat. 3391404, filed in 1959.

[VU 10] VU T., SUDALAIYANDI S., DOOGHABADI M. *et al.*, "Continuous-time CMOS quantizer for ultra-wideband applications", *Proceedings of IEEE International Symposium on Circuits and System*, pp. 3757–3760, 2010.

[WIE 91] WIESBECK W., KAHNY D., "Single reference, three target calibration and error correction for monostatic, polarimetric free space measurements", *Proceedings of the IEEE*, vol. 79, pp. 1551–1558, 1991.

[ZEB 16] ZEBRA TECHNOLOGIES, available at: www.zebra.com, 2016.

[ZHA 06] ZHANG L., RODRIGUEZ S., TENHUNEN H. *et al.*, "An innovative fully printable RFID technology based on high speed time-domain reflections", *Conference on High Density Microsystem Design and Packaging and Component Failure Analysis*, pp. 166–170, 2006.

[ZHE 08] ZHENG L., RODRIGUEZ S., ZHANG L. *et al.*, "Design and implementation of a fully reconfigurable chipless RFID tag using inkjet printing technology", *IEEE International Symposium on Circuits and Systems*, pp. 1524–1527, 2008.

[ZHU 11] ZHUO Z., Impulse radio UWB for the internet-of-things: a study on UHF/UWB hybrid solution, PhD thesis, University KTH, Sweden, 2011.

[ZWI 13] ZWICK T., WIESBECK W., TIMMERMANN J. *et al.*, *Ultra-Wideband RF System Engineering*, Cambridge University Press, 2013.

Index

Printed in the United States
By Bookmasters